OGDEN'S CIGARETTES.
A FIELD HOUSE.
OGDEN'S CIGARETTES.
ORCHARD POULTRY HOUSE.
OGDEN'S CIGARETTES.
SUBURBAN POULTRY HOUSE.
OGDEN'S CIGARETTES.
OYSTER SHELL
FLINT GRIT
GRIT & SHELL BOX.
OGDEN'S CIGARETTES.

COCKEREL OR PULLET?

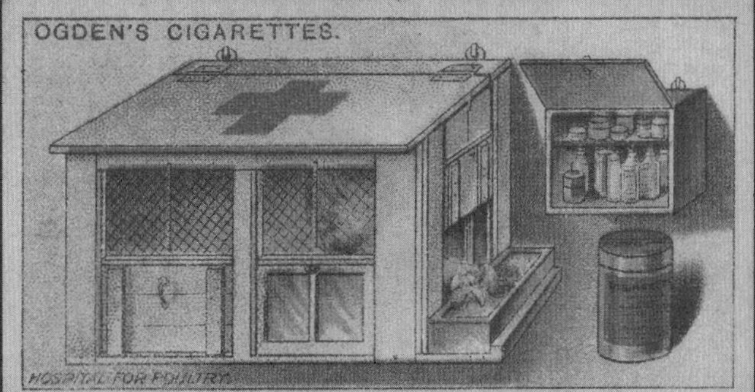
OGDEN'S CIGARETTES.
HOSPITAL FOR POULTRY.

A CHICKEN IN EVERY YARD

SURLY
SURLY

A CHICKEN IN EVERY YARD

THE URBAN FARM STORE'S GUIDE TO CHICKEN KEEPING

Robert and Hannah Litt

TEN SPEED PRESS
Berkeley

Published in the United States by
Ten Speed Press, an imprint of the
Crown Publishing Group, a division
of Random House, Inc., New York.
www.crownpublishing.com
www.tenspeed.com
Ten Speed Press and the Ten Speed
Press colophon are registered
trademarks of Random House, Inc.

Front cover illustration by Chris Hotz, Nemo Design.
Illustrations on pages iii, vi, vii, 6, 26, 29, 31, 32, 34, 36, and 38–41 by Diane Jacky (www.dianejacky.com).
Photos on pages ii and 173 © 2007–2010 by Patrick Barber.
Illustrations on page xii, "Keep 'Em Flying," and page 96, "Grow Food," from the Victory Garden of Tomorrow by Joe Wirtheim.
Photos on pages v, xi, 7, 14, 24, 42, 53, 57, 64, 84, 101, 117, 129, 130, and 172 by Dan Johnson.
Illustration on page 10 by Jeff Bartel, Nemo Design.
Photos on pages 18, 27, 46, 60, 67, 74, 82, 88, 95, 97, 99, 104, 109, 110, 118, 125, 128, 138, 142, 155, 169, and 176 by Robert Litt.
Illustration on page 20 by Ben Gurnsey, Nemo Design.
Photo on page 68 by Joshua Guuerra and Christina Pucci.
Photos on pages 70 and 114 by Lubosh Cech © LuboshCech.com.
Photos on pages 72 and 79 by TheGardenCoop.com.
Photos on page 81 by Cynthia Schubert.
Coop diagrams on pages 91-93 by Robert Litt.
Illustration on page 132 by Tim Kirkpatrick.
Illustration on page 164 by Leo Battersby, Nemo Design.
"Anatomy of an amniotic egg" on page 167 by Frank Horst.

Library of Congress Cataloging-in-Publication Data
Litt, Robert.
A chicken in every yard : the Urban Farm Store's guide to chicken keeping / by Robert and Hannah Litt. — 1st ed.
p. cm.
Summary: "This nuts-and-bolts guide to selecting and raising backyard chickens offers everything a first-time keeper needs to know, from the experts at the Urban Farm Store in Portland, Oregon"—Provided by publisher.
Includes index.
1. Chickens. I. Litt, Hannah. II. Title.
SF487.L775 2011
636.5—dc22

2010028185

ISBN 978-1-58008-582-3

Printed in China

Design by Katy Brown

10 9 8 7 6 5 4 3 2 1

First Edition

We dedicate this book to our neighbor Ann Nixon, who loved our chickens.

CONTENTS

TOP BAR
BEE HIVE
LOCALLY MADE
$250
URBAN FARM
STORE
MASON JARS
for PRESERVING
Ball

PREFACE

IT'S NOT UNUSUAL for couples to meet and later discover that they share an artistic passion, work in the same field, or a have a common religious tradition. For us, it was chicken keeping. Hannah had the most experience, having been around chickens since she was a child. I had been a backyard flock-farmer for only a year or so, but it had already become my favorite hobby. During our first summer together, we enjoyed the company of my small flock of two hens, knowing that we would soon add to it. Indeed, by the following spring our flock had expanded to eight birds, and the presence of several hatchery catalogs on the coffee table suggested that it would soon expand further.

That summer, we did not have a chicken-themed wedding, but we did get a lot of chicken-related gifts (thanks, Mom!). Hannah had recently found work in her true calling, nurse-midwifery, and I had just gone back to school for an advanced degree in agricultural science. While Hannah brought home the bacon, I studied the production of bacon.

During the course of my studies I became interested in how animals are fed and what effect these agricultural systems have on their health and, ultimately, our own. I remember looking at the label on my bag of organic chicken feed, trying to decipher what it contained: terms like "processed grain by-products" and "plant protein products" seemed a little vague and possibly sinister to me. I did some research and found that feed (that's "food" to you and me) was labeled in this way to give big mills the flexibility to formulate their feed from the cheapest ingredients available at any given time. Although this might be acceptable for most farmers, I suspected that I was not alone in wanting something more consistent and potentially healthier for my birds.

To test my theory, I first approached the folks on my local online chicken forum. I asked them whether they would be interested in buying feed that was grown mostly in our region and had quality ingredients that read more like a recipe than a list of commodities. With their encouragement, I commissioned my first 1-ton batch of "Portland Layer" (that is, feed for laying hens) from a local feed mill that did custom blending.

Following the lead of so many start-up businesses, I began selling this odd new feed from my garage. I often wondered what the neighbors thought of the mysterious cash transactions for enormous, unlabeled white bags. As the traffic around our house increased, Hannah suggested that perhaps a proper storefront would be more suitable for my venture than the driveway of an old Portland neighborhood.

FUN FACT

There are more chickens than humans on earth.

I opened the Urban Farm Store in the spring of 2009 with a few bags of feed, a couple of cases of suet, and high hopes that I could transform a ramshackle former recording studio into a feed store. Hannah worked alongside me on her days off, and together we steered the little shop through a season of chick buying and selling, gradually increasing our customer base for the feed and helping others discover the joys of chicken keeping along the way.

From my perch behind the counter at this little store, I researched and answered people's questions about their chickens for hours each day. Here I met folks who had elevated the chicken to full-fledged pet status—stopping just short of dressing them in tiny chicken sweaters (though I wouldn't be surprised if a few of them gave in to the temptation). I met other folks only interested in maximizing their egg production, and even a few who wanted me to teach them how to "harvest" their birds for meat.

I've learned a lot about chickens from my customers and still more from my wife. Together we'd like to share the best of this information with you.

ACKNOWLEDGMENTS

WE WOULD LIKE TO THANK our chickens: Rosie, Yankee, Wissahickon, Ginger, Butternut, Buttercup, and Tweedy. We don't know how they pulled it off, but somehow our lives seem to revolve around them.

Thanks also to our families and friends for tremendous support and encouragement along the way. From recipes and recipe testing to suggestions for book titles, they had our backs. Iris, Jorjan, Annie, Mari, David, Stuart, Dale, Evan, Jonas, Bill, and Martha are dearly loved and much appreciated.

Hannah would like to thank her friends and childhood neighbors, the Hazzard family, to whom she traces the original inspiration for chicken keeping and backyard livestock keeping in general.

We'd also like to thank Tonya and Chris for sharing their extensive chicken experience with us since we opened the Urban Farm Store. I'm sure they can see their advice throughout this book.

Thanks to all the people at the Urban Farm Store for everything you do.

Last, but not least, we'd like to acknowledge the newest member of our family. Abigail, although you probably had no idea what was happening outside of mommy's belly, we want you to know that you were our greatest inspiration to finish the book on time.

1

WHY KEEP CHICKENS?

WHEN WE TELL SOMEONE that we have seven chickens living in our backyard, there is usually a brief silence, followed by a curious "Why?" Hannah will counter, "Why not? They're great pets—charming, and useful, too." Indeed, which of your other pets provide you with wholesome, protein-rich food and ask so little in return? Does your dog keep the bug population at bay in your backyard? Will the cat mow the lawn for you?

Besides these practical considerations, we think chickens are just plain fun to be around! We derive so much delight from observing the antics of our hens that we would likely keep them even if they did not lay. Indeed, many of our customers are surprised and pleased to discover that chickens have a certain kind of grace and can be truly beautiful. Because of this, Hannah likes to call them "mobile lawn ornaments." Robert gains the deepest satisfaction from watching them methodically graze the lawn or scratch through the compost seeking hidden bugs. As he cares for these long-domesticated animals, he feels a deep connection to the land and to a not-too-distant agrarian past. Chickens are at once so useful, colorful, and entertaining that we can no longer imagine our lives without them. Best of all, our hearts swell each time we see the look of absolute delight and fascination in the eyes of children when they see their first live chicken bounding across the lawn toward them, or when they hold a warm, fuzzy chick. Children seem to know intuitively that chickens are special creatures with a lot to teach us all about nature and ourselves.

QUALITY OF LIFE AND SUSTAINABILITY

The chicken is best known as the producer of that humble staple food, the egg. Although the egg is familiar to all, we can safely say that you have never truly experienced its full potential until you have eaten one laid by one of your own hens and cracked open and cooked while still warm. Not only will its deeply colored yolk and firm white taste richer and more flavorful than you ever imagined an egg could, but this special egg will provide an immense satisfaction earned from your role in its production.

Backyard hens enjoy an enviable lifestyle compared to their commercial counterparts. Even so-called free-range hens often suffer from crowding in immense, climate-controlled barns; your hens, on the other hand, will enjoy a relatively carefree life full of affection, quality food, and fresh air. Couple that with the opportunity to preserve older, heritage breeds and their unique traits, and you can see why Robert likes to say that keeping a home flock truly "allows chickens to be real chickens."

Another thing we love about producing our own eggs in the backyard is that they don't have to travel to reach our table. When it comes to commercial eggs, organic included, the fragile orbs are typically deeply chilled to preserve them and then trucked many miles to your plate. Even eggs with distant "use by" dates usually were laid weeks, if not months before and are no longer anywhere near their best. Keeping a few hens in the backyard is a great way to conserve resources by eating food that's grown so close at hand that the "local" label is more appropriately replaced by "homegrown." This is one reason many so-called locavores have especially embraced backyard animal husbandry. Only chickens and a few other small animals can provide high-quality protein in the city or

Origin of Domestic Chickens

Chickens were domesticated from a wild red jungle fowl (*Gallus gallus*) at least eight thousand years ago in Southeastern Asia. By 2000 BCE they had reached the Indus River valley and soon thereafter spread to Europe, the Mediterranean, and Africa. The domesticated chicken was established throughout Polynesia by about three thousand years ago, and there's evidence that they reached the Americas several centuries prior to European contact. Chickens became an important food source wherever they were introduced because of their small size and ease of care.

Like other animals that humans have domesticated, chickens differ in at least two relevant ways from their wild ancestors. First, domesticated chickens are much less aggressive than their wild counterparts. Although this makes them easier to handle, it also means that they require more protection from predators. Domesticated birds are also much more prolific producers of meat and eggs than their wild cousins. This results in higher feed requirements and the need for further human support. The result: a red jungle fowl can take care of itself, whereas a domesticated chicken is almost totally dependent on humans for food and protection.

suburbs. It's easy to see where your food comes from—and even easier to get to know the farmers when they eat breakfast with you!

Some backyard chicken keepers do raise birds for meat, and if you eat chicken, this is certainly a good way to ensure that the birds you are eating had a happy life and ate well while they were alive. This will not be a major focus of this book, however, because the vast majority of backyard chicken keepers regard their chickens as pets and find it unsettling—if not outright upsetting—to consider eating them. We feel that this is a choice for the chicken keeper to make; we take no position either way. If you are interested in leaning more about harvesting your birds for meat, visit www.urbanfarmstore.com to find more information and links to useful resources.

Finally, it's important to consider that the eggs (and meat) from your backyard buddies will contain a more optimal balance of nutrients than their store-bought rivals. As we'll discuss in chapter 8, recent findings suggest that eggs from small, pasture-raised flocks (like yours) are lower in cholesterol and have a healthier ratio of omega-3 and omega-6 fatty acids than even organic, free-range eggs from the store. Remarkably, the nutritional comparison also showed that these eggs were higher in several important vitamins and minerals. Eggs that taste great, are laid by happy hens, and are good for your health—what could be better than that? We'll go into a lot more detail about eggs in the chapter we've creatively titled "Eggs."

AFFECTION

Although few chickens show their owners the blatant dog or cat sort of love, many of our customers report that they have a mutually affectionate relationship with their hens. Some birds do seem to genuinely enjoy human company and will seek physical contact. Our store manager, Sharon, has just this sort of relationship with most of her birds. At the store she regularly picks up and handles our Crested Polish hen, Muppet, who shows her appreciation for the affection by cooing and burrowing into Sharon's arms. At home, she has birds that are sometimes invited indoors to watch TV on her shoulders.

Although it's debatable how much chickens enjoy demonstrative humans, it is clear that their owners often become very emotionally connected to them. Our customers joyfully share tales of first eggs, humorous antics, and moments of concern for their birds' well-being, clearly indicating that strong emotional bonds form with these animals as with other pets.

EDUCATIONAL VALUE

Some of you may have grown up on a farm and experienced firsthand the many life lessons that caring for livestock can provide for young children. For the rest of us, our childhood experiences with farm animals were probably limited to petting zoos and fuzzy-paged children's books. If you had some direct experience, keeping a few hens in the yard is a terrific way to pass on what you know. If not, it's a great opportunity to learn together as a family project. Either way, it's clear that kids seem to get a special kind of enjoyment from chickens. The sense of responsibility for and connection with the natural world that children develop from caring for any animal is immeasurably valuable. If that animal happens to offer eggs in exchange for the care, the bond created is even more powerful. When a child picks up his or her first egg from the nest, this incredibly nutritious food becomes very intriguing indeed. You might even find that a kiddo who used to turn up her nose at scrambled eggs suddenly begins to clamor for them.

We have seen many children come into the store who have clearly been studying their chickens. They can recite the exact names of breeds they own, often pronouncing the sometimes complex words better than their parents. Some of them have even written essays and book reports on chicken keeping, astounding us with their encyclopedic knowledge. Others have brought their prized birds into their classrooms for show-and-tell events—much to the delight of their classmates, no doubt. Whether learning proper holding techniques, hen-housekeeping procedures, or feeding routines, these kids really seem to enjoy finding out all there is to know about their unusual and compelling pets.

Many local schools in our area now have chicken coops because of their tremendous educational value. Of course, students and teachers have been raising chicks and even hatching eggs in classrooms for years. Usually, the lesson plan will center on biological aspects of the chicken life cycle, embryology, and other developmental topics. These days, our teacher customers have also been raising and keeping the chicks into adulthood to teach their students about where food comes from, nutrition, and sustainability.

If the chickens become ill, or when they eventually pass on, there are even deeper lessons to deliver to young ones. For adults and children alike, there is undeniable sorrow associated with the loss of any pet, and chickens are no exception. However, unlike dogs and cats, who regularly live over a decade, chickens will usually live only four years or so (up to eight under ideal circumstances). This fact means that you and your family will need to face the transient nature of life relatively often. Although this could be seen as a liability, we prefer to use the occasion of a hen's passing as a time to reflect on the unique give-and-take these creatures share with us and ponder our own place in the natural cycle of life. That said, we have also helped several parents hurriedly replace chicks that have died suddenly, slipping them into the brooding box before the kids come home from school to discover the loss. You can play it either way.

AESTHETIC VALUE

Another reason to keep chickens is that they are just plain appealing to look at. Chickens have long been appreciated for their feather colors and attractive forms, traits that have earned them a place in the art of cultures around the globe. Beyond their appearance, chickens have a unique gait that seems at once comical and graceful. They probe their environment with scratching and pecking motions that we find endlessly fascinating. A flock of chickens adds such a beautiful sense of movement to your garden that this alone makes them a worthwhile addition to your yard.

Silkie

To accentuate this beauty, most of our customers choose a variety of different breeds, rather than raising a flock of all one type. In our breeds chapter (chapter 3), you will discover that chickens not only come in different colors but also have different feather patterns, sizes, and even body shapes. Chickens are available in a startling variety of breeds, from delicately laced-patterned Wyandottes (as seen on page xi) to Silkies, which look more like fur-bearing mammals than birds. Then there are the truly bizarre breeds: Naked Necks resembling vultures, frizzled breeds with heavily twisted and crimped feathers, and feather-footed, orblike Cochins, to name a few. In fact, there are enough types of chickens out there to keep the backyard flock owner researching and pursuing new varieties to collect year after year. Or, as many people do, they may decide to specialize in a favorite one, even going so far as to breed them or ordering eggs to hatch themselves.

Before Hannah got her first batch of chickens, she spent months poring over catalogs, books, and online resources, deciding which breeds were the best and most beautiful. Aesthetics should be considered, because your birds will be a long-term feature of your backyard, but we think that each breed and individual has its merits. That said, we would like tell you that we have never had an ugly chicken, but that would be a lie. Her name was Springsteen (because she was a Jersey Giant), but she was neither giant nor did she have the attractive, glossy black plumage characteristic of the breed. Her dull gray-black feathers were unevenly distributed over her body, leaving bare patches that became tough and reddened from exposure. She was an eyesore—and rather unappetizing to look at when we ate dinner on the patio. We probably would have kept her despite her pitiful appearance, if it were not for her bullying of the other hens in our flock. Her saving grace was that she was a prolific layer of enormous eggs, which made it possible to pawn her off on a farmer friend of ours. Call us shallow, but for us it's not only about the eggs.

FUN AND HUMOR

Regardless of their motivations for crossing roads, chickens are just plain funny. It seems that some of our customers make visits to the store just to relate the latest antics of their hens. One woman told us her chickens love to entertain themselves by swinging on her hammock. They hop up on the edge and flutter their wings, trying to keep their balance as the hammock swings wildly to and fro. Another customer's hen likes to head-butt her cat. Our own wily chickens tag-team unsuspecting guests at backyard barbecues: one chicken will approach the victim and provide a distraction while her partner in crime hops up and neatly snags the desired food right out of the victim's

Stunt chicken!

hand. This has happened enough times now that we are convinced these chickens truly are in cahoots. Finally, we occasionally have our own little Easter egg hunt when a clever chicken of ours decides to hide her eggs in a new location. Once we found a cache of twelve eggs inside a pole bean trellis. Another time she had us completely stumped— until we went out to the composter and nearly dumped out the compost pail onto a little collection of eggs that she had been hiding there! It's hard to know whether our hens are intentionally having fun with us, but it sure seems like it.

GARDEN HELPERS

Another great virtue of a backyard flock is their eagerness to pitch in around the garden. They are expert insect hunters and can make a substantial dent in the population of unwanted garden pests, such as slugs and tomato-chomping caterpillars, while having little effect on populations of most beneficial ones, like ladybugs. Though earthworms are hardly pests, they are generally abundant, and your birds will also avail themselves of a worm whenever possible, transforming them into a quick, high-protein snack, and eventually into an egg. Now that's true alchemy! Chickens are not just hunters; fallen fruit and other potentially wasted garden produce will be quickly consumed as well. Beyond scavenging, they also eat green plants with a methodical intensity. A small flock of hens grazing a lawn are tapping into a valuable nutritional resource, all the while helping to reduce your mowing chores. In fact, the American lawn of today looks the way it does because it's an idealization of a grazed, pastoral landscape. Returning animals to the pasture, even on a tiny scale, just makes sense to us.

This gusto for plant browsing also has its drawbacks. Chickens are big fans of leafy ornamental plants like hostas, and if given free rein they will mow them down faster than you can say "locusts!" Similarly, chickens will gladly devour your lettuces, spinach, kale, and other green leafy vegetables. It's possible to grow a plant palette that avoids becoming expensive chicken food; for example, they will generally avoid herbs and other strongly scented plants. But you will likely want to fence off or otherwise protect your prized ornamentals and veggies before you sort out what your flock will and won't want to eat. (For further information, see "Chickens in the Garden," page 124.)

Chickens will also scratch the earth in search of tasty tidbits, making shallow holes and possibly exposing the roots of plants. Although this activity is usually harmless, you should keep it in mind when planning your garden-protection scheme. You can harness these nibbling and digging tendencies to help you in the garden as well. Chickens can be confined to a pen or other structure and made to concentrate their digging and foliage chomping in a small area, effectively clearing it for future planting.

Perhaps the greatest assist that your birds will provide in the garden will come not from the beak end of the bird, but from the other. Your flock's pooping practices will produce enough soiled bedding to supercharge your compost pile, and they will spread nutrients wherever they are allowed to roam. Contrary to popular belief, widely dispersed chicken poop in this form will not burn your plants (though it is advisable to hot-compost larger amounts of poultry waste to ensure that any potentially harmful organisms are eliminated). Indeed, we have noticed a marked improvement in the health of plants in the backyard, where the chickens live, compared to the front yard, from which they are excluded. Chickens and gardens seem to love each other's company, and we think that you will love the combination, too.

Chicken Family Tree

Chickens share an ancestor with an unexpected animal: the *Tyrannosaurus rex*! In 2008, Harvard researchers compared proteins from a nearly seventy-million-year-old T-rex with those of modern creatures. The result? Chickens and ostriches were the closest matches, meaning they are distantly related to the fearsome dinosaur. They even share the same wishbone (furcula) shape in their skeletons. You will find that your chickens are not nearly as fierce as their distant cousins, although they will occasionally surprise you with their speed and cunning when foraging in your yard.

INDEPENDENCE
Over Easy

2

PLANNING: ARE WE READY FOR THIS?

WE HOPE THAT the opening chapter has convinced you that chickens are a lot of fun and can make an enriching addition to your life. However, keeping a flock in your yard is not all fun and games. We tell our friends and customers that it's important to do some homework before bringing home a basket full of peeps. Chickens, like other pets, require a commitment of time, energy, and a little money. Chicks will need to be kept indoors for their first few weeks and tended to several times a day; adults will need daily care and a safe place outdoors to live (although we do know of one flock that spent a year in an attic during a remodel!). And before you build that coop, don't forget to check with your city or neighborhood association to confirm that it's legal to keep chickens at your home and to learn what rules need to be followed. Lastly, remember to be considerate of your neighbors and inform them of your plans—bribing them with promises of eggs as needed.

LIFESTYLE CONSIDERATIONS

At the outset, you'd be wise to consider whether this unusual pet would be a good fit for your lifestyle. Judging by our diverse range of customers, the answer is usually "yes." Chickens are really quite low-maintenance animals—we like to say that they take about as much time and energy as a cat does, once your chicken-keeping setup is complete. This initial setup does require planning, a short burst of energy, and, above all else, commitment. Most of this initial energy will go into coop building (or purchasing and setup), which, if done right, should be largely a one-time investment of time and money. After this, chickens in a flock of two to ten birds should require an average of five to ten minutes a day, mostly for feeding, watering, and egg collection, punctuated by an occasional coop cleaning.

You'll also need to judge whether the cost of keeping chickens is prohibitive for you. As we will discuss in a moment, coops can be expensive to build or buy, and you will need to invest in some durable equipment at different stages of your flock's life. Don't forget that feed and litter will also be recurring expenses. Compared to your other pets, chickens are actually pretty reasonable, but they do cost more than just chicken feed!

As we mentioned earlier, keeping chickens also means being present to do daily chores, even if they are easy. If you spend much of your time traveling away from home, you may want to postpone actualizing your chicken-keeping dreams until your life settles down a bit. That said, the lure of free eggs will have neighbors clamoring for the privilege of caring for your chickens when you are away on short trips. We even know a number of neighbors who have entered into chicken keeping as a joint venture. This is a smart way to go (assuming that your neighbor is someone you want to deal with regularly), because costs, responsibilities, and eggs are all shared, which makes it much easier to work chickens into your life.

Other folks think that chickens are not for them because they work long hours or have busy social lives. If this is the case, you will need to make some slight modifications to your routine and some major changes to your chicken-keeping setup, but it can be done. Chickens do not need to be babysat during the day; they are perfectly capable of taking care of themselves if they have a secure environment and flock-mates to keep them company. However, if you regularly return home long after sundown, think about how you can get your chickens safely shut in their coop in your absence. Happily, chickens will go back into their coop at dusk on their own—they just need someone to shut and secure the door behind them to keep predators out. We have totally unpredictable schedules ourselves, so we chose to install a rather expensive automatic door that operates on a timer. For us it was well worth the price for the peace of mind it affords us when we're running late. There are, of course, other less expensive options. Perhaps there is a neighborhood kid who will close the coop for you every day after school in return for eggs or a small fee.

We have also heard people fret because they always neglect their houseplants to death and wonder what would happen to a chicken in their care. We tell folks like this to not let their lack of a green thumb stop them, because chickens are much better than a ficus at telling you (loudly and vigorously) what they need. A houseplant fades into the background, whereas chickens lure you every day with the promise of fresh eggs and delightful antics. You won't forget them, we promise!

Other prospective chicken keepers may be daunted by stories of people having to deal with alarming chicken health problems that sound too complicated to treat. Like any pet, chickens do have occasional health issues, but one of our goals with this book is to empower you to prevent or address most of them easily yourself. Indeed, most backyard chicken keepers we know—and we know hundreds—never have to address a major chicken crisis at all. It's also reassuring to know that more vets are welcoming chickens into their practices and can offer help with anything that you don't feel capable of dealing with yourself.

Having children should not stop you, either. Chickens are fascinating for kids, and they will introduce basic concepts of food sources, animal care, and responsibility. As we mentioned in the previous chapter, children seem to be delighted by chickens and are eager to learn about them. If you play your cards right, you might be able to rope the kids into doing some of the chores for you.

SPACE REQUIREMENTS

In addition to time and energy, you will also need an appropriate spot in your yard for your birds. Those with no outdoor living space will be hard-pressed to find a way to house their chickens—a sixth-story apartment is a difficult place to imagine accommodating a successful flock. Some imaginative chicken owners keep their hens in diapers running around the house, but we don't recommend this approach. As Robert likes to say, "Chickens need a place to express their chicken-ness." At a minimum, you will need an area in your backyard that is large enough to accommodate a coop (the enclosed chicken house, also called a henhouse) and a run (the securely fenced pen adjoining the coop).

We hesitate to quote exact figures for how many square feet to provide per chicken, because most studies that have been done on this subject are for commercial laying operations, not for home flocks. We have found that the per-chicken requirement is a whole lot different if you have three chickens than if you have a hundred. As we often say, backyard chicken keeping is an art rather than a science.

A spacious backyard coop

When customers ask what they'll need for a typical backyard flock of three birds (which is a manageable number for a beginner and is usually the legal limit without a specil permit), we tell them that they'll need an area of at least 4 by 8 feet—more (say, 8 by 10 feet) if the birds will be mostly confined to the space all day. We have much more to say about space and building requirements in chapter 5, but we're throwing out some numbers here to help you decide whether your back (or front) yard can accommodate a small flock. In addition to space for the coop and run, plan on allowing your birds access to the whole yard to forage and frolic for even a few hours a week if at all possible. They will reward you for this with better health and egg production, not to mention lots of cheap entertainment.

In addition to available space, your backyard should meet the basic environmental needs for this type of animal. For instance, chickens require both sun and shade at various times, and they will not thrive in places that are either excessively wet or overly dry. Most yards meet these minimum requirements or can be easily modified to do so. However, if you live on a houseboat, you might want to think twice before building a coop in that leaky old rowboat. Or just get ducks instead!

WHAT'S THIS GONNA COST ME, ANYWAY?

The cost of keeping chickens depends largely on the choices you make at the outset, and the single most important decision you will make is how to approach coop building. At the Urban Farm Store, we have seen our customers' choices run the gamut from practically free to thousands of dollars. For instance, several intrepid folks we know have built their coops for under a hundred dollars by using almost entirely reclaimed lumber and hardware. Although this may not be practical for most of us, using some reclaimed building materials will help to keep both cost and environmental impact low. Most customers build a solid but modest coop for a couple hundred bucks and construct it themselves as a weekend project (or five weekends, if you are as slow as we were). Others hire a contractor to build the chicken palace of their dreams. We even know one customer who spent hundreds of dollars building a coop that resembles Hogwarts from the Harry Potter books. They call it "Henwarts."

Of course, before you build the coop, you will also be on the hook for the chicks and the equipment for raising them. This will total about $75 to $150, depending on a few factors we will cover later in the book. After the initial layout for the coop, chicks, and their supplies, along with a few other durable accessories for the adults (up to an additional $100), the cost of chicken maintenance is essentially tied to the cost of their feed and litter (that is, the bedding that helps keep the coop clean). We figure this at about $5 to $10 per chicken per month, depending on whether you go for a homegrown, basic, or premium diet and how often you clean out the coop.

If you think you'll be saving money on eggs, try dividing the cost of your coop and other expenses by the average cost of a dozen eggs. You'll quickly surmise that, for most of us, it takes many years of chicken keeping to recoup the investments. Though you may not soon be saving money on your grocery bill, we think the cost is more than justified in other ways (see chapter 1). In the final analysis, your hens will always be pets, not money-savers. We do, however, know one adorable young chicken keeper who sells homegrown eggs to her neighbors to supplement her allowance. It would be hard to resist that sales pitch, to be sure!

Keep in mind that getting things set up correctly from the start will be the most thrifty approach in the long run. For instance, trying to save money by building a flimsy coop will lead to unhealthy chickens, possible losses to predators, and frequent repairs—neither efficient nor economical. Similarly, we recommend that you invest in the appropriate feeding and watering accessories. We've known many people who've tried to keep using their chick feeders for their full-grown chickens, only to eventually realize that the cost of the wasted feed (from the chickens repeatedly kicking over the feeder) would have paid for a durable adult feeder. Likewise, we encourage you to seek out and use the best feed that you can buy. Quality feed will likely improve your birds' health and laying—not to mention the nutritional quality of the eggs—enough to easily justify the outlay.

FUN FACT

*Alektorophobia is the **fear of chickens.***

AM I ALLOWED TO KEEP CHICKENS?

Not too long ago, a few chickens pecking around the front yard of a house in an American city was a common sight. It was a matter of simple practicality; people had a vegetable garden and some chickens because these were two of the easiest ways to grow their own food—and they still are. After WWII, the rise of the supermarket began a sudden and pronounced shift away from self-sufficiency. At about the same time, the emergence of the suburbs signaled the end of most home chicken keeping, as restrictive community rules were drawn up that limited pet choices to cats and dogs. This anti-chicken bias was particularly evident in former farming communities wishing to shed their rural image.

These days we are witnessing a widespread reversal of these attitudes. Many cities and their surrounding suburbs are revising their ordinances to reclassify chickens as pets, not livestock. The result? Chickens have become downright trendy. Even sophisticated New York City is fascinated with the chicken-keeping movement. As proof, our friends and family regularly send us articles clipped from the *New York Times*, *New Yorker* magazine, and other highbrow publications about chickens and the people who keep them (including Martha Stewart). Across the pond in England, backyard chicken keeping is an established hobby that has been mushrooming in popularity lately. According to the *Guardian*, there are currently over half a million UK households keeping chickens!

Here in Portland, Oregon, where chickens have been allowed more or less continuously since the 1800s, they are practically an institution. There is a popular online chicken forum and an annual backyard coops tour, and our little store has been featured on the news several times. Of course, we dearly hope that the current resurgence of love for chickens is not just a flash in the pan. Unlike most trends, chickens are downright practical, so we think that they are here to stay.

When trying to figure out whether your home town is hip to chickens, we suggest that you start by consulting the online resources that list various cities and their current laws regarding chicken keeping. We will not attempt to print a list of cities and laws here—it would be out of date by the time this book made it into your hands. See the Resources for a list of some of the more comprehensive websites listing chicken laws and regulations.

Even if chickens are allowed in your area, consider your neighbors before you take the chicken plunge. It's in your best interest to notify them in advance about your plans. Tell them how you plan to keep the chickens confined to your yard (with a fence) and keep the odor nonexistent (cleaning the coop regularly and keeping a reasonable number of birds) and the noise down (by not keeping any roosters—which many cities don't allow anyway). By involving your neighbors in your plans and educating them in advance, you will avoid future conflict with them. You can also try persuading them with the promise of eggs!

If you know in advance that one of your neighbors is extra finicky, it would be wise to position your coop as far away from their property as possible (25 feet seems to be a safe distance), and anticipate any complaints they might have in advance. In some cities (Portland is one), there are some regulations in place that protect both neighbors and chicken keepers. If the county receives a complaint, an inspector will arrive at your home (announced) and inspect your setup. They are usually helpful and are often a great resource for helping to resolve disputes.

Authors' hens

Hannah's personal experience with this scenario was actually a very positive one. The inspector was compassionate and even complimentary about the chicken coop's construction and the rest of her setup. However, Hannah says that if she had known in advance that the particular neighbor in question was anti-chicken, she probably would have been able to avoid the inspection altogether by selecting a different spot for the coop. She also could have provided the neighbor reassurance that there would be no roosters to keep him awake—which ended up being his underlying concern.

It is also important to keep in mind that, even if your city is amenable, your homeowners or condo association may have other ideas. Although they may not have the force of law behind them, these institutions can make your life difficult and hit you where it really hurts: the wallet. Similarly, renters would be well advised to consult their landlords before acquiring chickens or building a coop. A customer recently approached me to offer us a small flock that they could no longer keep because their landlord had decided to sell their rental house and asked them to leave. As you might imagine, they were having trouble finding a suitable place to live on short notice where they would be allowed to keep their birds.

OUR TWO CENTS

Lastly, it's a good idea to seek out the advice of experienced chicken owners. After all, no one knows what it takes to raise chickens better than a seasoned chicken keeper. At most feed stores that sell chicks the salespeople are happy to spend a few minutes sharing basic chicken-keeping tips. There may also be classes in your area that will help you get started. At our store, we offer classes on a regular basis both for chicken newbies and for experts who want to learn more.

Be wary of asking people with a serious farming background about keeping chickens. They have usually kept them on a large scale and may have an entirely different outlook on these creatures than you will. We once overheard a customer asking a farmer friend of ours how he suggested carrying their chickens to the vet. He replied, "I just tie their feet together and toss them into a burlap sack!"

SHOW YOUR CHILDREN
WHERE FOOD REALLY
COMES FROM

3

BREEDS: PICKIN' CHICKENS

YOU'VE DECIDED TO take the plunge and get yourself a few hens for the backyard. The spouse has been convinced, the rules for your city have been checked, and the planning has begun. Now for the really fun part: deciding which of the dozens of available breeds are right for your tastes and your unique situation. There are birds that lay more eggs than you can eat each week, others robed in gaudy feathers, and even a few that may eat the slugs but leave your lettuce alone (Hannah believes these exist in legend only).

One trick that you may find to be useful at this point is to make a few informed but arbitrary decisions such as "I will only consider heritage breeds" or "I really like eggs and need a selection of top laying hens." To aid in this process, this chapter includes our lists of top birds in several major categories and in-depth breed profiles for our favorites. Don't assume you'll choose just one type. A well-rounded flock may consist of several of these birds, each one special in its own way. We suggest that you mix them up; a variety of breeds makes for an exciting visual experience and helps to highlight interesting breed differences. Don't worry about how they'll get along: the warnings about breed incompatibilities you may read about in other sources mostly apply to large flocks, not to backyard pets.

CHICKS OR YOUNG ADULTS?

When it comes to starting your flock, your first decision will likely be whether to go with chicks or more mature birds. Older birds are appealing to those who feel that chicks are too fragile or time-consuming for them and want to start with something lower-maintenance. It's important to remember, however, that by selecting older birds you will usually have a more limited selection of breeds to choose from. You will also miss out on the opportunity to socialize the chickens yourself, which helps them adapt to your lifestyle.

Pullets

Female hens less than one year old are called pullets, though this term is generally reserved for "teenage" birds from two to six months old that are not yet laying (six-month-old hens are sold as "at point of lay"). Pullets are often available from local farms and individuals who either raise them to sell or need to relocate extra birds. If you are looking for layers, we recommend purchasing pullets rather than older hens, because egg production is best in the first two years. Older hens do make lovely pets, however.

If you choose to start with older birds, there are a couple of things to watch out for. First, make sure that the hens are actually hens! If you do not want to buy a rooster (which will be true for most backyard chicken keepers), you can spot the hens by looking for flat or rounded "saddle" feathers on the back just above the tail (see page 26 for more on this). On roosters these tend to be pointy at the end. This is the most reliable way that we have found to differentiate between the sexes early on, before the telltale "cock-a-doodle-doo!"

Next, check the bird for any obvious health problems. Hold the bird and examine the skin beneath the feathers for parasites or sores. See "Proper Chicken Handling," page 127. Inspect around the chicken's bottom and particularly under her wings. Look for hens with clear eyes and nostrils, red combs, lustrous feathers, and a solid build (somewhat heavy). Avoid birds that sneeze, wheeze, or otherwise seem to be sick in any way. Also check them for white, aphid-like bugs under the scales of their legs; if you see any, decline the bird politely (and wash your hands!).

While examining the legs, you will also want to check for blanching. As a hen ages, her legs will often lighten from a deep yellow, blue, or gray to a lighter shade. Unless you are looking for a pet only, you will want to avoid buying older birds, because they will not lay as well as a younger bird, if they lay at all.

Finally, check out the facilities the chickens are kept in and the health of their flock-mates. If they are being kept in crowded, unclean conditions, be aware that this sort of setup breeds disease. Likewise, if other chickens in the flock look ill, there is a good chance that your prospective chicken has been exposed to whatever nastiness is afflicting the others.

Chicks

We strongly suggest that you begin your flock with chicks. Beyond the vastly superior selection you will have to choose from, it's far more fun for most people to raise these adorable, cheeping fuzzballs. They are entertaining to watch and satisfying to care for, and they will be more relaxed with being handled by humans later in life if they get used to receiving regular attention as youngsters.

When it comes to finding chicks, you will usually have only two alternatives: mail ordering directly from hatcheries or purchasing from local stores. As you might expect, there are pros and cons to each. Mail order is a good option because of the massive selection of breeds you will have to choose from. There are hatcheries listed in the back of this book that stock dozens of breeds of sexed (90 to 95 percent female) birds. These will be available for shipping within a week or so of order placement during the peak months (February through July) of chick season. They may not always have all of the breeds you

FUN FACT

*A **breed of chicken** called the Silkie has furlike feathers, **iridescent earlobes**, and black bones.*

want at the same time, but the larger hatcheries do a good job of offering alternative breeds or shipping dates to satisfy your needs.

The big drawback of ordering directly from a hatchery is that they either ship in twenty-five-chick minimum batches (so they keep each other warm) or will charge you a lot to ship a few birds (with included warming packs). Also, although you will pay for only the birds that survive, shipping may be very stressful for the chicks and a few may arrive dead or weakened. Assuming that most of the twenty-five birds make it to your door, unless you're really ready to house them all, you will likely need to find homes for the extras by advertising or sharing an order with a friend.

Local farm stores will not have the wide selection that mail-order hatcheries will, but we suggest that you start with them anyway, for several reasons: you will be able to get just the number of birds you need, you will not have to deal with any stressed or dead chicks, and you will be able to ask lots of questions while you are there.

Two-week-old chick

We suggest calling ahead to determine what your local store has in stock and what they will be getting in soon. Because you have about a three-week window to mix chicks of different ages before they become incompatibly different in size, knowing what's arriving in the near future in addition to what they have on hand will allow you to research which breeds will be best for you and decide on an appealing mix. We often get asked whether different breeds get along. As we noted earlier, they do. Differences in age and levels of dominance within the flock matter far more than feather color. That said, you should think twice before mixing birds of vastly different size or breed temperament in larger flocks, as this is where differences become more pronounced.

In addition to asking about breeds, we also want you to ask the following questions, and if you don't get a "yes" to all of them, consider looking elsewhere. First, ask whether the birds are "sexed." Sexing is a screening that occurs at the hatchery to separate the girls from the boys. Because most cities and towns do not allow roosters (males), you will need to purchase only pullets (females). Chicks that are not sorted by sex will be referred to as "straight run."

The sexing process is complicated and not something that can be done at the store, except for a few special hybrid breeds called sex-links. These chickens have been bred in such a way that all females will be one color and all males another. Even when done properly, traditional sexing is only partially accurate, so be sure to ask if they have any policy on returning chickens that turn out to be roosters. Some will allow you to drop them off (usually to an uncertain fate) after they show their "true feathers" of masculinity at about eight to twelve weeks. Some of these roosters will join country flocks, and some will find their way to a dinner table. If you are especially concerned about their fate, consider hand-delivering the birds to either a farm or sanctuary. Either way, don't expect to be compensated for your time and expenses.

You should also ask whether the chicks have been vaccinated. Chicks are commonly vaccinated for a nasty thing called Marek's disease. In most cases you're better off with vaccinated chicks, though some hatcheries with excellent breed stock have never seen the need. There is also now a vaccine available for coccidia, a tiny parasite that causes all manner of trouble in backyard flocks. If this vaccine is available, by all means accept it, but do not expect all hatcheries to offer it at this time. Luckily, there are other measures you can take to protect your flock from coccidiosis. which we explore in chapters 4 and 7.

Finally, we want you to ask whether chicks of different breeds are kept separate from each other in the store. We have found that when many different types of chicks are kept together in the same brooder, the chances of an employee fishing out the breeds you really want are pretty low. It is acceptable, and usually unavoidable, for some very distinct-looking chicks to be kept together. This is fine as long as the staff are reasonably well trained and the store stocks a manageable number of breeds.

Roosters

The process of sorting males from females occurs immediately after hatching. This work can be done only by highly trained experts, so it occurs exclusively at the hatcheries. Even these experts make mistakes, but the process is remarkably effective, averaging 90 to 95 percent accuracy. This means that you have about a one in ten chance of getting a young rooster chick when you should be getting a female.

We get asked every day to help customers decide whether they have a rooster in their young flock. I first ask about the chick's age, because before about eight weeks it's impossible to be sure of anything. Next we tell them that aggressive or bossy behavior, large combs or wattles, body size, tail feathers, and stance do not indicate much of anything. Then we tell them about what really matters: saddle feathers. These are the feathers that develop where the lower back meets the base of the tail. Again, in hens they are flat or rounded; in roosters they are pointy at the end. Of course, the best indicator is a good, old-fashioned "cock-a-doodle-do!"

COMMON BREED PROFILES

In the following section we provide descriptions of ten common breeds, lists of recommended breeds in various categories, and a comprehensive chart summarizing these and other well-known types. This chapter is not for breeders, collectors, or poultry historians; there are many other books that cover these subjects in greater detail. Rather, this chapter is intended to tell backyard chicken keepers what they really need to know about chickens, based on our understanding, research, and conversations at the store with chicken owners. Therefore, we have focused more on personality, ease of care, and egg-laying ability than on proper feather color and breed perfection standards.

Bantams

It is also important to note that many breeds are available in "bantam" forms—half-sized versions of chicken breeds that are kept more for novelty and pet quality than for laying. These pygmy poultry are cute, but you should proceed with caution. Because of their tiny size after hatching, they are generally available only as "straight run" or unsorted by sex. This means that about 50 percent of the chicks will be roosters. It's also important to consider that these smaller-bodied birds are good fliers and adept at escaping the confines of a backyard, making them difficult to keep. They may go easier on your garden, however.

Personality

Many people want to know about the personality or temperament of different breeds. Although we agree that there are some general differences and tendencies among birds of a feather, we have found that socialization, or the way you raise your birds, makes far more of an impact on the behavior of individual birds than their heritage. If you handle the birds gently and frequently from the time they are chicks, the adults of even the most independent breeds will tolerate (or

Store hen, Muppet

even enjoy) human contact. Conversely, individual hens of a "friendly" breed that have no experience with humans will be shy and avoid interaction. Don't necessarily avoid breeds because you hear that they are high-strung—just know that you will need to make a bit more effort when raising them and accept the possibility that they will keep their distance from you. Keep in mind that breeds listed as "docile" are not necessarily going to be friendlier to humans. This simply means that they are less likely to bully other chickens in the flock—and they might be more susceptible to bullying themselves.

Meet the Breeds

Without further ado, here's the scoop on the top ten breeds you'll want to know about.

Ameraucana

Get out the ham, because this bird lays green eggs! Actually, they can also be blue, or bluish green. Just like other colorful eggs, on the inside they are the same as all the rest. Beyond the fantastic color of the eggshells, this bird is one of our favorites because of its sweet personality. A must-have!

Personality: This bird is an absolute pleasure to have around. Easygoing, yet seldom bullied, this bird is easy to socialize and will likely become a favorite of adults and children alike.

Size and Description: Small to medium-sized, sleek body, tight feathers, "bearded" appearance typical. Hens 5.5 pounds.

Color Variations: No standardized colors; they can be white, cream, brown, red, partridge, buff, black, or a combination of colors.

Eggs: Strong seasonal layer of large blue, green, or occasionally pink/brown eggs. Summer: about five eggs per week. Winter: about one or two (if any) eggs per week. Not prone to going broody (that is, to stop laying in an attempt to hatch eggs; see chapter 7 for more information). Some will stop laying altogether in cold, dark winters.

Care: A relatively low-maintenance bird to keep. Seems to be slightly more apt to have "crossbeak" (see chapter 7).

History: Originally bred from a chicken from South America called the Araucana, which had a genetic problem, this bird was perfected through crossbreeding in the 1970s. This makes it one of the newest breeds, but it fits right in with a heritage flock due to its rarity and unusual eggs.

Notes: The staff at the store have voted this the most underrated breed. We always suggest including this bird even if you are getting only a few chicks. The egg color alone would recommend it, but the personality is the icing on the cake.

Australorp

A great choice for any backyard flock, they are great layers, beautiful, mellow, and will accept a variety of environments with aplomb.

Personality: Australorps are not prima donnas. They will be happy kept in a run or roaming about your yard. They are unlikely to bully other chickens, but are shy enough that they probably will not be your most petlike bird either.

Size and Description: Medium-sized, heavily feathered. Hens 6.5 pounds.

Color Variations: Black, but very shiny, with purple and green iridescence visible in sunlight.

Eggs: Hard-to-beat layers of medium brown eggs. An Australorp holds the record for most eggs laid in a single year (364!). Yours are unlikely to match that, but they certainly will perform well. Summer: about six eggs per week. Winter: about four or five eggs per week. Australorps can blame their Orpington heritage (see Orpington listing) for their tendency to go broody.

Care: For a bird bred in a warm climate, these do great in cold areas also. No specific health concerns for this breed.

History: The Aussies got ahold of some Orpingtons from England, and when they were done mixing in some of the best layers around, they ended up with this superb specimen. The American Livestock Breeds Conservancy (ALBC) considers the Australorp to be "recovering" in its scale that rates the danger of extinction of each breed (ratings range from "critical" to "threatened," "watch," "recovering," and "study").

Notes: The Australorp is an unsung heroine. She doesn't attract much attention, but once you have one, start noticing her quiet beauty, and get used to how regularly she lays, your flock will feel incomplete without one.

Brahma

Brahmas are sure to be appreciated by anyone who has ever kept one (like us). Although other breeds get most of the attention, these birds quietly go about their lives in their majestic way. An excellent choice for kids because of their calm nature and sturdy constitution.

Personality: We said "quietly" in the introduction for a reason—Brahmas, along with Delawares, are among the most quiet hens around. They are also unquestionably one of the most docile, calm, and easily tamed. Their large size helps to reduce the bullying by other members of the flock experienced by other docile breeds.

Size and Description: Large, heavily feathered, with fancy feathered feet. Hens 8 pounds, occasionally up to 9 pounds.

Color Variations: Light, dark, and buff; the dark variety has amazing tweed-like penciling on its feathers.

Eggs: Surprisingly good layers of brown eggs. Three or four eggs per week, all year long. However, these gals have a biological clock that tells them to brood relatively frequently.

Care: A robust bird with few problems.

History: These hardy birds originated in India, where they ranged from the hot coastal plains to the Himalayas. They were brought to the United States in

the mid-1800s. They were originally used for both meat and eggs, but today they are more of an ornamental bird. These birds have been given ALBC "watch" status; your owning them will contribute to their comeback.

Notes: We love Brahmas for their lovely feathers and sweet personality. Trust us, you need one or more of these birds in your flock!

Maran

Marans owe their increasing popularity to the unique color of their eggs. In our own flock we have had only positive experiences with them. We named our first Maran "Kiwi" after the tailless national bird of New Zealand, because she remained tailless well into her adolescence. Adorable.

Personality: Variable, depending on the origins of the Maran in question. We have not heard personality-related complaints from our customers.

Size and Description: Large, thickly feathered birds. Hens 7 pounds.

Color Variations: Black Copper, Silver Cuckoo, Golden Cuckoo, White, Wheaten, Birchen, and Blue. The common silver cuckoo type looks like the more common Barred Plymouth Rock, but the Maran has white legs; the Plymouth Rock's are yellow.

Eggs: Decent layers of large, deep chocolate brown to copper colored eggs. Summer: three or four eggs per week. Winter: two or three eggs per week.

Maran

Orpington

Care: Health attributes also vary by strain, and it will be very difficult to determine the breeding stock that led to the chicken in question.

History: Originally from France, and developed in the early 1900s, the breed now has standards in both France and England. Both types have been imported into the United States, where they are not yet APA-recognized. Since then, the distinct strains have been interbred, leading to unusual variation within the United States population when it comes to breed characteristics.

Notes: If you want a rainbow of colors in your carton, seriously consider a Maran. There are other dark egg layers, but these are by far the easiest to acquire.

Orpington

Big and puffy, with an almost comical appearance, the Orpington is a very popular chicken for the home flock. Their soft plumage makes them attractive to handle. This increased socialization may account for their reputation as a friendly chicken. Fast maturation is a plus.

Personality: The Orpington will usually get along well with other members of the flock, but may sometimes be bullied by more dominant birds. We have found them to be generally uninterested in human contact, but our customers report a mixed bag: some find them to be extremely personable, others not at all. All agree, however, that they are not aggressive. Free ranges and tolerates confinement very well.

Size and Description: Large body, loose feathers, massive appearance. Hens up to 8 pounds.

Color Variations: Most commonly seen in buff yellow with light legs and a single, medium comb. Less commonly available in white, black, or (slate) blue.

Eggs: Moderate layer of large-sized, light-colored brown eggs. Summer: about four eggs per week. Winter: about three or four eggs per week. Prone to going broody, which will mean fewer eggs overall. These birds are known for steady cold-weather laying.

Care: A relatively low-maintenance bird to keep. No health problems specific to the breed. Tendency to go broody is something of a problem.

History: A classic British breed from the village of Orpington in County Kent, the Orpington was originally introduced in 1886 in its black form. From the start it was a dual-purpose meat-and-eggs breed (like most at that time) that was an especially good layer of brown eggs in winter. Later, the buff form was introduced; it has since come to dominate. No longer used commercially, the Orpington is listed as "recovering" by the ALBC, thanks in large part to the breed's popularity in backyards across the UK and the United States, particularly the buff color variation.

Notes: Buff Orpington chicks are the classic little yellow fuzzballs and are quite irresistible—this may account for some of their popularity. The Austral-orp, discussed earlier in this list, is considered an Australian Orpington (thus "Austral-orp").

Plymouth Rock

Another all-time classic chicken breed, this was one of those first two chickens that got Robert hooked! Unbeatable for lively backyard antics, hardy in rough weather, and very attractive. Great layers that are very fast to mature, usually starting to lay three to four weeks before other heritage breeds.

Personality: A real team player, these birds are fantastic in mixed flocks. Docile and easy to socialize, but seldom bullied by other birds. Great foragers that can find lots to eat in your yard.

Size and Description: Large, thick body, tight feathers. Hens 7.5 pounds.

Plymouth Rock

Polish

Color Variations: Barred (alternate markings of two colors, usually black and white) is by far the most common coloration. Yellow legs and a single, medium comb. Also available in Black, (slate) Blue, Buff, Columbian, Partridge, Silver Penciled, and White.

Eggs: Strong layer of large-sized, medium-colored brown eggs. Summer: about five eggs per week. Winter: about three or four eggs per week. Not prone to going broody.

Care: A relatively low-maintenance bird to keep. No health problems specific to the breed. Withstands foul weather well due to tight feathering.

History: This is a true American original, originally bred in the mid-1800s. Long a popular breed for family farms, in the 1950s Plymouth Rocks lost out to modern specialty breeds and became "threatened." Now classified as "recovering" thanks to free-range poultry producers of both meat and eggs. Also a very popular backyard breed.

Notes: Children seem to delight in the appearance of these "zebra chickens." This bird deserves a place in most flocks.

Polish

Polish are a fancy breed. Barely recognizable as chickens to the casual viewer, they are an exotic departure from ordinary chicken-ness.

Fancy Feathers

Columbian coloring is white with black highlights on the tail and tips of wings, and black with white penciling around the neck. This coloring came to be when a chicken of this description was exhibited at the World's Fair (otherwise known as the Columbian Exposition) in Chicago in 1893. Partridge coloring is a reddish brown base coloring with multilayered darker brown or black penciling on each feather—presumably because it resembles the delicate coloring of some partridges.

Personality: Our customers report that the personalities of their Polish range from flighty to lap-chicken. Those with very large crests may behave strangely because their vision is obscured. Unlikely to be at the top of the pecking order in a mixed flock, these chickens can be subject to pestering from flock-mates who are intrigued by the unique crest. As a solution, a few folks have purely Polish flocks.

Size and Description: Small birds with an upright posture. "Crested" head, which means they look like they are wearing a headdress. Hens 4 pounds.

Color Variations: Black, White, Blue, Golden, Silver, Buff, also available with "laced" feathering, bearded varieties, and those with a contrasting crest.

Eggs: Lay small white eggs. Were once known as good layers, but as the breeding focus has been on enhancing appearance, Polish are variable in how often they lay. Very unlikely to go broody.

Care: Most health problems that the Polish experience are caused by the very feature that makes them unique: that heavily feathered head. The feathers block clear sight, so your Polish may be more prone to injury. They can be pecked at by other chickens. Apparently, problems also arise if these headdresses remain wet for long periods of time, or if ice can accumulate in frigid weather. However tempting, *do not* trim the crest. This may cause serious harm.

History: Originated hundreds of years ago in Europe, but apparently are not from Poland. Most likely from the Netherlands.

Notes: We have never had a Polish in our home flock, because we are somewhat egg-centric. However, we have had several in our store flock, and they always steal the show. Kids can't get enough of them.

Rhode Island

Renowned for its prolific egg laying, hardy constitution, and perky personality, the Rhode Island is a backyard chicken of legendary status. At the store we find it easy to recommend Rhode Islands to beginners and experienced chicken keepers alike. Our oldest and most favored hen, Rosy, is a Rhode Island Red.

Personality: They have been described as "friendly," "perky," "curious," and our favorite, "docile yet regal." What does that mean? They're not aggressive and can easily be socialized to enjoy handling. Intelligent and largely unafraid of people, these hens will eagerly tag along with you in the yard. Although they prefer to range freely, they will tolerate being confined to the run if needed. May be slightly bossy to other chickens in the flock, but this is seldom a problem. We love them!

Size and Description: Medium to large, blocky body. Hens 6.5 pounds.

Color Variations: Red with yellow legs and a single, relatively large comb. There is also a white version, but it's far less common.

Eggs: Great layer of large-sized, medium-colored brown eggs. Summer: about six eggs per week. Winter: about two to four eggs per week. Rarely go broody but do make good mothers (for those chicken keepers who can keep a rooster).

Care: A relatively low-maintenance bird to keep. No health problems specific to the breed. In fact, we have found them to be especially robust—our matri-

Rhode Island

Sussex

arch hen Rosy recovered from a serious dog attack in a couple of weeks and looks great!

History: Developed in the mid-1800s from crosses of Red Malay Game, Leghorn, and Asiatic stock. Aptly named, this breed hails from Rhode Island, where it is the official state bird. A dual-purpose type originally bred for both meat and egg production, this breed is no longer used in large commercial farms. Considered "recovering" by the ALBC, the Rhode Island is making a comeback in free-range farming operations, where their hardiness and early maturation are appreciated.

Notes: Not to be confused with New Hampshire Reds, Production Reds, or Red Sex Links—all fine birds, but distinct from the classic Rhode Island Red. If possible, seek out Rhode Island chicks designated as "true" or "dark," which should help to ensure that you are getting the genuine breed.

Sussex

Sussex are a dual-purpose breed, prized as meat birds because of their heavy-breasted build. However, they are also fantastic layers, which gives them a solid place in the backyard chicken movement.

Personality: We adore our Sussex hens. They are quite curious and will be the first chicken to come running when they see you. They will be right at your feet whenever they get the chance, so they are very easy to pick up and pet—a delight because they are so soft. They also seem to do equally well confined or free-range.

Size and Description: Large birds with a rounded body shape. Heavily feathered and unbelievably soft. Hens 7 pounds.

Color Variations: Speckled, red, and light (speckled is by far the most available in the United States, and the most beautiful in our opinion).

Eggs: Excellent year-round layers of three or five light brown eggs per week. The only thing that stops these egg-laying machines is a period of going broody that happens about once a year.

Care: Very strong and hardy. No special problems.

History: Originally from Sussex, England, where they were bred in the early 1800s.

Wyandotte (White, Gold Laced, Silver Laced)

Notes: Sussex remain more popular in England and Canada than in the United States, but we aim to change that. This fantastic breed should be considered for any backyard flock.

Wyandotte

These beauties are hard to resist! There is something very pleasing about the round body shape, and they come in a spectacular variety of colors. Combine this with their egg-producing ability, and they really fit the bill for functional lawn ornaments.

Personality: Most sources identify Wyandottes as docile and friendly. We find that in a mixed-breed flock they are "assertive" with lower-ranking chickens at any opportunity. That said, we continue to add more to our flock because we just can't resist their other attributes.

Size and Description: Medium-sized, rounded body shape, loose feathers. Hens 6.5 pounds.

Color Variations: Silver Laced and Gold Laced are the most popular (and very beautiful); also found in White, Buff, Partridge, Silver Penciled, Blue Laced Red, and Columbian.

Eggs: Solid layer of large-sized, light to medium brown eggs. Good choice for year-round laying. Summer: about four or five eggs per week. Winter: about three eggs per week. It's unusual for this bird to go broody.

Care: An excellent selection for colder climates. They lay well in the winter, and their rose comb resists frostbite. Generally healthy and trouble-free.

History: Another true American breed, the Silver Laced variety was first developed in New York in the late 1800s. Created to be a dual-purpose breed for both eggs and meat, they are now quite popular with small-scale chicken keepers. The wide variety of plumage options can be attributed to breeders' interest in how beautifully their body shape shows off the feathers. Listed as "recovering" by the ALBC.

Notes: If one of your priorities is to have an attractive flock, but you don't want to sacrifice much in the way of eggs, you need to get a Wyandotte or two.

The Lists

These brief lists will help you make a decision about which chickens to get, based on a variety of criteria.

Ten Best-Selling Breeds at the Store

1. Rhode Island—A favorite for egg productivity and perky personality
2. Orpington—Solid winter layer, attractive, and docile
3. Plymouth Rock—Great layer with decorative coloration
4. Ameraucana—Green eggs and a sweet disposition
5. Wyandotte—Very attractive and a good winter layer
6. Australorp—Great layer, docile personality, and attractive plumage
7. Brahma—Big, fluffy bird that's easily handled
8. Maran—Dark brown to copper eggs and great looks

9. Polish—Dramatic crest makes a statement
10. Sussex—All-around solid performer and attractive

Most Prolific Egg Layers (in ideal conditions)

1. Leghorn—Up to 325 white eggs a year
2. Sex Link (and other hybrids)—Up to 300 brown eggs a year
3. Rhode Island—Up to 275 brown eggs a year
4. Australorp—Up to 275 brown eggs a year (record-holder for most eggs laid in a year: 364)
5. Sussex—Up to 250 brown eggs a year

Best with Kids

1. Brahma—Docile yet large enough to withstand a little rough play
2. Cochin—Very docile and fluffy
3. Orpington—Large and docile
4. Ameraucana—Underrated as a pet
5. Plymouth Rock—Easily socialized

Heritage Breeds to Preserve

1. Buckeye—Midwest classic now nearly extinct
2. Java—One of the oldest American breeds
3. Dominique—Another old American breed, being replaced by Plymouth Rocks

4. Delaware—A dual-purpose bird (meat and eggs) that's nearly been lost to time
5. Sussex—British dual-purpose breed—old school!

Most Unusual Appearance

1. Silkie—Looks like a mammal with fur!
2. Naked Neck—Lack of neck feathers is cute to some, bizarre to others
3. Frizzled Cochin—Curled and contorted feathers on a round body make it unique
4. Polish—Regal crests make an elegant impression
5. Belgian D'Uccle—Cute, tiny, and heavily feather-footed

Unusual Colored Eggs

1. Ameraucana or Araucana—Green to blue eggs
2. Maran—Dark brown to copper eggs
3. Barnevelder—Dark brown eggs
4. Welsummer—Dark brown to copper eggs
5. Penedesenca—Dark reddish brown eggs

For information on more breeds, we recommend visiting Henderson's Breed Chart at http://www.ithaca.edu/staff/jhenderson/chooks/chooks.html. We think you'll agree that the incredible diversity of breeds is a big part of why keeping chickens is so compelling.

4

CHICKS: PARENTING YOUR PEEPS

FOR MANY OF US, our only contact with baby chicks has come from the basket of adorable, fluffy chicks we were given on Easter morning. We snuggled with them and allowed them to roam the halls of our doll houses—until they mysteriously disappeared a few days later. Neither we nor our well-intentioned parents understood that chicks simply require more to sustain them than a basket full of artificial grass and bits from a chocolate eggshell. Like babies of most species, chicks need special care for their first several weeks of life. In addition to the classes taught at our store, we give our customers a two-page chick care handout with each batch of chicks they purchase. Beyond this simplified instruction manual, there is much more to learn about caring for these cuties to ensure better survival and a more enjoyable experience. We're confident that the well-tested advice we have gleaned from thousands of customers and our own experiences at the store will give you the tools you'll need to help your young birds thrive.

PREPARATION

We have already led you through the range of lifestyle, legal, and logistical considerations relating to your prospective chicken ownership. After reading about the startling diversity of chicken types and breeds, you may have already decided which are most appropriate for your backyard flock. Now it's time to start preparing your home for the arrival of your feathery new friends. This can be likened to creating a nursery for an expected human baby, though not nearly as expensive or time-consuming. Like preparing a nursery, planning for this step now will greatly ease chick care and maintenance later on, and make sure your "babies" are comfortable from the moment they are placed into their new home.

Preparation begins with creating a snug and safe place for them to spend the first several weeks of their development. It is helpful to understand that in the wild (and in some farm situations), baby birds rely on their mothers to care for them in a process known as brooding: hens hatch their eggs and then warm, protect, and feed the young birds until they are independent enough to strut their

Chick Checklist:

- ❑ **1 quart waterer/fount, top and base**—invest in the 1-gallon size if you plan to have more than five chicks
- ❑ **1 quart feeder and base**—buy the long, multislot feeders if you will have more than five chicks
- ❑ **Litter**—pine shavings or other low-aroma, nonpulpy material made for pet bedding
- ❑ **Feed**—25 pounds of chick feed, usually medicated (yes, they will eat that much!). Never use adult feed.
- ❑ **Thermometer**—useful to monitor temperature inside the brooder
- ❑ **Heat lamp and bulb**—invest in a good lamp with ceramic base, heavy-duty cord, and bulb guard. Get a 250-watt red Pyrex bulb.
- ❑ **Brooder box**—*large* cardboard box or tote (ideally should be already in place before birds come home)

stuff. For most backyard flocks, the process of brooding begins at a hatchery and continues at home in an artificial "brooder." You and your family get to play the role of mother hen.

During their first six to eight weeks of life, your young birds have four fundamental needs: warmth, nourishment, security, and sanitation. We like to say that when all of these requirements are being met, the birds are thriving in a low-stress environment. We have found that sad chick stories usually result from one or more stressing factors, such as incorrect temperatures, rough handling, improper nutrition, or dirty drinking water. Keep in mind that some of these problems can be caused by excessive care or "overdoing it" as well as by neglect. Never fear; we will help you strike the correct balance that will give your chicks the best possible chance to thrive.

We call the place where your chicks will spend their critical period of development the "brooder" or "brooder box." This actually consists of several elements, but primarily refers to a container of sufficient size coupled with a heat lamp (sometimes called a "brooder lamp"). For the box, most people simply use a large cardboard box; others build wooden structures or repurpose wash basins or other suitable containers. You are limited only by your imagination and the functional requirements for a brooder, summarized as follows.

Accommodations

For a typical batch of chicks (three to five birds), we recommend that the brooder box ideally be at least 2 feet by 4 feet and 2 feet deep, though you can go smaller initially if you are able to closely regulate the brooder's temperature as detailed below. Why do we like the box to be so big? A sizable brooder will allow for the rampant growth and activity of the chicks, which will each reach the size of a small football by the time they can go outside. Just as important, a large interior volume will allow you to create both a warm and a cool zone inside the brooder. These warm and cool zones will provide distinct areas for setting up the feeder and waterer as well as areas that allow the chicks to self-regulate their temperature up or down as needed by moving around.

Our simplest (and least expensive) suggestion for a brooder box is to use two large cardboard moving boxes; strap them together with packaging tape on the outside, cut out the middle box walls to make one large box, and fold the flaps inward for strength. Refrigerator boxes and garment boxes laid on their sides with one side removed to make a top opening also make excellent brooders. Some sources recommend against cardboard because it can get wet and could conceivably catch fire if your heat source became damaged and came into direct contact with it for extended periods. However, if you keep your litter clean and dry and take reasonable precautions with your heat source (recommended in any situation), this should not be a problem. Cardboard boxes are convenient, cheap, recyclable, and perfectly adequate for the average family planning to raise five chicks or fewer.

Brooder box at the store

If you want a sturdier, more permanent box, you can easily build a simple rectangular box out of plywood. For instance, the wooden coop depicted later in this book (see page 88) can also be used as a brooder for young birds and later moved outside. Another popular choice is to use a large container of some kind that you plan to use in the future for another purpose. For instance, you could use a large metal stock tank (what horses and other large livestock drink from), then later in the summer you could drag it outside and use it for your vegetable container garden. Others use very large plastic totes. These are convenient because they are sturdy, reusable, and easy to find at major chain stores. The middle of the lid can be cut out and replaced with hardware cloth to provide ventilation without compromising protection. If using a plastic tote, be careful with the positioning of the heat lamp to avoid melting the sides of the brooder.

Perhaps the most creative and simplest approach to a brooder box can be found in a spare bathroom. My customers have taught me that bathtubs are ideal in size and depth for up to eight large chicks. They love how easily the tub can be cleaned, they find the shower curtain useful for keeping active chicks (and dust) contained inside, and they report that a shower head makes a perfect place to hang a brooder lamp (if you can get it at the right height—see the "Warmth" section). When using a bathtub, it is important to cover the drain hole to prevent litter from clogging it and to keep smaller chicks from falling in. Also, keep the toilet lid down! It is hard to predict when the first adventurous chick will be able to escape the confines of the tub. Although she may be feisty, don't count on her also having the judgment to avoid drowning hazards.

Covers

Whichever type of brooder you choose, be sure to use a sturdy cover that will allow unimpeded air flow yet protect the chicks within. Cats, in particular, will reach into your brooder to extract the chicks if given a chance. Even if cats are not a threat in your home, the chicks will eventually want to fly out, leading to frantic chick hunts and unexpected chick-poop encounters.

For this purpose we recommend a fine wire screening material called hardware cloth (see chapter 5 for details). Ideally, you would make a wood frame and stretch the wire material tightly for maximum durability and protection. You can also simply mold the material around the edges of the brooder to conform to its shape. Feel free to experiment with other materials; just be sure that your selection is fireproof, will not melt, and does not have any sharp parts that could injure the birds.

Warmth

As mentioned already, chicks need an additional source of warmth until their feathers become developed and their body size increases to the point they can withstand cooler weather. In a more natural situation, this warmth is provided by the mother hen's body heat, trapped in her downy feathers and, in some cases, through direct transfer of heat from her skin via a "brood patch." With pet chicks, heat is typically provided by the brooder lamp hung above or clipped to the side of the brooder box.

When it comes to brooder lamps of any type, safety is paramount. Hot bulbs, cardboard, and litter can combine to create a potential fire hazard. Always make sure that the lamp is of the proper type for a high-wattage heating bulb (with a ceramic base, heavy-duty cord, and protective grill) and that it is secured carefully above or clipped securely onto the side of the brooder so that it can't be dislodged accidentally. These lamps can be easily obtained at a good feed store or most hardware stores. Beware of lamps with a plastic base; these will not stand up to the heat generated and could melt or smolder. It is also important to use a heat lamp bulb made of Pyrex to prevent shattering in case it comes in contact with moisture while hot.

We also strongly recommend the use of a 250-watt red heat bulb to facilitate proper chick sleep patterns and avoid pecking injuries. With a white light, chicks have trouble sleeping appropriately and can become "overstimulated," for lack of a better term, which can in turn trigger aggressive behavior. The cozy red glow from a red bulb will allow your chicks enough light to get around and stay warm, without keeping them up too much. This type of lamp will also keep visual contrast low and prevent potential problems with pecking at the feathers, toes, or wounds of their brooder-mates.

We have found that using a large brooder box greatly simplifies temperature regulation by giving the birds areas for warming and cooling themselves as needed without frequent lamp adjustment. This is also helpful if you have chicks that vary in age (and thus warmth requirements) in the same box. The chicks' chosen "hangout" area in their brooder will tell you a lot about how comfortable they are in the environment you are providing. For instance, if your chicks spend a lot of time clustered in the circle of heat under the lamp, they are too chilly and your lamp should be hung lower and closer. Conversely, if you find that the chicks are avoiding the lamp like the plague, they are overheated and you should raise it a bit. In a perfectly heated brooder, the chicks will be distributed throughout the floor at various times of the day, happily going about their business.

You should locate the lamp on one side of the brooder box to create your warm zone, leaving the other end of the brooder to serve as the cool zone. This means either hanging the lamp above the brooder off-center or clipping it to one end of the brooder (be careful—the clamp can sag). In either configuration, it is very important that you leave enough room between the hot bulb and the side of the brooder. As the chicks grow older and their need for heat decreases, you may still wish to adjust the lamp height (or angle, if clipped) once or twice in a large box to ensure that the chicks are using the whole space and not avoiding an area directly under the bulb because it's too hot.

In smaller brooder boxes (anything less than our recommended size), it is especially important to adjust the height of the lamp to regulate the temperature as the chicks grow more warmth-retaining feathers. This usually involves starting the brooder temperature at 95°F for newly hatched birds (as measured with a thermometer), and then raising the lamp incrementally so that the temperature decreases by 5 degrees per week of age until the birds go outside at a final temperature of about 70°F. As you can see, this is much more complicated than simply using a larger box for your brooder.

In either method, the bulb should start at about 24 to 30 inches above the floor of the brooder at normal room temperatures to achieve a zone of 90 to 95°F below it in the warmest area of the brooder. This is the lamp height that works in most situations, but if your brooder is in a cool basement, for instance, the lamp will need to be lower, and if it is in your very cozy living room, it may need to be higher. That said, we would not recommend putting the brooder in any areas of extreme temperatures. Choosing a location for your brooder that has relatively consistent temperature will help tremendously in allowing you to regulate the temperature inside the brooder. For example, if you try to keep your brooder in an unheated shed, and the temperature is 70°F during the day but 40°F at night, you are going to have to do some serious heat lamp adjusting every morning and evening.

FUN FACT

A chicken's heart beats 280 to 315 times a minute. Its body temperature is 102°F–103°F.

Drafts can also be a problem, as breezes can quickly chill lightly protected chicks. For this reason, brooders need to be indoors or in a well-constructed outbuilding, and the sides of the brooder need to be high enough to keep out even slight breezes. This is also one reason why we prefer a heat lamp to a heater that blows heated air at the chicks.

On very warm days in rooms without air conditioning (like our store), you may need to turn off the lamps for some or all of the day to keep the temperature in the brooder at the correct level. Follow the guidelines for the age-to-temperature relationship and turn the lights on and off accordingly. Again, a thermometer is a very handy tool to keep near the chicks.

Litter

This is the technical term for the stuff you strew on the bottom of your brooder to cushion your chicks and absorb moisture. Believe it or not, the choice of material is an amazingly controversial topic among chicken keepers. You can read five books on the topic and come away with five different vehemently stated recommendations. Our local chicken-related online forum regularly hosts lively debates on the subject, and there is never a consensus. Since our store opened, we have experimented with a number of materials and talked to hundreds of folks who have done the same, and we've formed some strong opinions of our own. However, with some exceptions, most of the choices out there are acceptable, and you can be guided by your personal preference.

One type of litter we *never* recommend for your brooder is cedar shavings. Cedar is far too aromatic for your chicks' immature lungs and can cause respiratory distress, which far outweighs any odor-control benefits. One of our most heart-wrenching stories came from a customer whose chicks began to gasp for breath in their bed of cedar shavings. After one chick passed away, their owner called us in a panic for advice. After hearing that she had been using cedar, we recommended that she immediately replace it with something else. She called us less than two hours later to report that her chicks had experienced a full recovery when moved onto paper towels. In an interesting twist to this story, this customer was an experienced chicken keeper and had raised her chicks in cedar chips in the past with no such problems. It may be this unpredictability that explains why some references still suggest cedar as an option. Perhaps it depends on the freshness of the shavings; nonetheless, it is one form of litter we strongly advise our customers to avoid.

In our quest for the ideal litter, we have tried many things at the store and at home. First we tried paper towels laid over the bottom of the brooder, based on worries that chicks would eat other types of litter when they were too young to know better. This got expensive and messy quickly, and we ditched the approach very early on when we saw no difference in chick health. We then tried using shredded junk mail and other scrap paper that we produced ourselves or swiped from the office. This was immensely appealing from an ecological and cost standpoint, but we soon discovered that paper turns to a stinky paste when it gets wet. We found that wood pellets were very absorptive and had a nice odor, but we later decided that they are too heavy (especially when saturated with chick waste), disintegrate into another mushy mess when wet, and are apparently appealing to the chicks as potential food when it gets to this state—gross! We already knew that plain newspaper laid down in layers would be too slick and this could lead to a condition called "splay leg" that could cripple your chick later in life, so we avoided that. We even considered using kitty litter, but immediately realized that it looks almost exactly like chick feed and would have plugged up our chicks' inner workings in an instant. It was suggested that we try using kitchen towels, old socks, or other bits of cloth. From what we gathered, we would need to change this form of litter daily, and we were not interested in adding to our already daunting laundry load.

After all of these experiments, we have found pine shavings to be the best option for our own needs, and for those of most of our customers. They are a relatively inexpensive, widely available by-product of an existing industry, are very fluffy and absorbent, provide good odor control, and are generally trouble-free. *Note:* When selecting pine shavings, look for a sifted product that has few small particles. These tiny bits are more easily eaten and can make an already dusty situation worse. You may have other options available to you locally, so feel free to experiment, within reason. In general, pine shavings seem to be the most reliable and safest option.

Feeders and Waterers

These are the final pieces of equipment that will make your chicks' brooder a functioning home. You can get these inexpensive items anywhere you can get chicks. The plastic ones will set you back a only a couple of dollars (metal ones are not much more) and are well worth the modest investment. They are specially designed to dispense water and feed throughout the day, so they will need to be cleaned and filled less often than other all-purpose containers you might try to use. The dispensing tray of the waterer is too shallow for a chick to drown in, which is an important concern as well. In addition, they are designed to make it difficult for your energetic little chicks to dig in (known as "scratching"), scatter, or otherwise mess up their food and water. We guarantee that they will spend the majority of their time in this pursuit.

Keeping their food and water clean and cool is one of the most important things you can do to maintain your chicks' good health. Whichever style of waterer and feeder you choose, you will want to locate it on the cooler side of the brooder box. Food tends to lose its vitamin content when warmed (though carbohydrates and protein are largely unaffected), and the water will quickly get slimy and possibly dangerous. If any stray bits of poop get in there (and they will), warm water will make harmful microbial growth occur much faster.

Security

The brooder must above all else be a safe place for the chicks. Safety starts with selecting a proper location for the brooder, where pets and young children will have only supervised access to the birds. There is nothing more deflating for a new chick owner than to come home to find that the family dog or cat has found (and "played with") the baby birds. Similarly, young children may overestimate the chicks' desire to play dress-up or fly with a parachute. When you cannot supervise interactions, it's best to prevent them entirely until both pets and kids have had plenty of supervised time to become familiar with their new family members. In the case of some pets (dogs with a strong chasing instinct, for instance), unsupervised interactions may never be a good idea. It's also important to consider that safety is a two-way street; chicks are generally not human disease carriers (see "Getting to Know You," page 60), but they are dusty and can be a little messy. Because of this, it's best not to locate the brooder in a child's room no matter how much he or she begs to sleep with the chickies.

With its limited access and steady temperature, a basement makes an ideal location to brood your chicks. If this is not possible, any room that can be locked is the next best place. It is possible to set up your brooder in the living room—just place it high above pets and children and be aware that you will need to be vigilant with your dusting. Have we mentioned the dust? To be honest, the quantity of dust a few chicks can produce is impressive. The combination of their rapidly shedding adorable chick fuzz and their vigorous foraging behavior will leave most flat surfaces in their general vicinity covered with a fine layer of white or gray dust.

BRINGING HOME YOUR BABIES

Once your brooder is set up, you have your supplies handy, and you know what to expect, it's time to get your chicks! The first thing to know is that chickens are social animals that live in flocks. If you were planning to buy a single chick to start with, think again. *You will need to purchase a minimum of two, and we suggest three.* Two birds will bond strongly to each other and do just fine. The problem is that if you lose one, for whatever reason, you will be down to one, which is unacceptable. A lone chick will try to bond with you, and although that sounds cute, it will cheep loudly and persistently if you are ever far from its side. This could make life stressful for both of you. It is almost as hard to ignore a crying chick as it is to resist attending to a crying baby. We had one customer insist on starting with one chick—it died and he came back a few days later for two new ones.

Two's company

As we detailed in the previous chapter, chicks are often purchased from mail-order hatcheries (see Resources); these offer a dizzying selection of breeds nearly year-round but usually require that you purchase a minimum of twenty-five birds per delivery (again, so that the chicks can keep each other warm en route). This is how Hannah obtained her very first chicks. She chose to do this so she could get the mix of birds she desired. However, she was then stuck with the task of finding homes for the remaining twenty birds.

A better option for the home chicken keeper is to purchase your chicks locally from a feed or specialty store. Many stores will let you pick out the individual chicks that you would like to take home with you. Take some time to check them out before making your choice. A healthy chick will be vigorously eating, drinking, and scratching about when it is not sleeping. Avoid any chick that is sluggish and unresponsive to the intrusion of a human hand. At our store, we identify listless chicks and move them to a "special care unit" in a backroom and treat them until we know they will be healthy enough to go home with a family, but most places don't do this. It is in your best interest to choose little ones that are frisky from the start.

That said, do not always go for the biggest chick you can find. Maybe that huge chick is just eating well or is older than the rest, but we generally find a higher percentage of eventual roosters among the largest chicks. Remember, "sexing" is not completely accurate, and 5 to 10 percent of the chicks will be boys (more, if your store is not careful with its ordering). Size will not always indicate a rooster, but just to play it safe it's better to avoid both the largest and the smallest (runts) and stick with the medium-sized ones. Finally, avoid chicks with any obvious deformities such as limping, a beak that doesn't seem to align correctly, or anything else about it that seems amiss.

Transporting Chicks

When the time comes to visit a store to buy your chicks and supplies, we suggest bringing your own container to bring the live birds home in. The name of the game here is keeping stress low. A covered box (with adequate ventilation) lined with paper towels will provide a much more pleasant journey for your tiny family than the paper bags most stores offer. As another measure to reduce stress, we advise that you shop for all of your supplies *first,* put them into your vehicle, and wait to actually collect the chicks until you are ready to leave the store. Gathering all of your supplies can be a long process, and chicks can become quite distressed in their frightening (and possibly chilly) new environment while they are waiting for you to finish shopping and checking out. Hopefully, you will have already set up the brooder before you bring the chicks home because—say it with us now—this too will reduce stress.

Upon leaving the store your new babies will be overwhelmed from being removed from their warm brooder and tossed into a cold box with unfamiliar neighbors. Because of this, it's critical to do what you can to keep things mellow on the way home. If it's cold outside, the vehicle should be kept warm (about 80°F if possible); if it's hot out, it should be cooled somewhat. Keep the radio and voices low and calm. This is definitely not a good time to allow children (or eager adults) to handle the chicks or even to open the box, as they may fly out, causing all sorts of mayhem. Drive safely, but hurry home and avoid stops along the way. Don't worry about your chicks being afraid of the dark in their little box. As you will discover when you become a more experienced chicken keeper, darkness and confinement are quite calming to poultry.

FIRST DAY AT HOME

When you get home, put your chicks directly into their prepared and prewarmed brooder. If you must set up a brooder first because you had to buy your supplies and chicks at the same time, keep the chicks in their box until the brooder is ready and keep the box in a warm, but not hot, place. We strongly recommend avoiding the temptation to pick up, play with, or pet your chicks on their first day home. Stress is the primary cause of a common difficulty called pasting up (more on this later), which is why we are so emphatic about minimizing stress. You can sit quietly and watch them discover their new environment, which we find endlessly entertaining. After about twelve to twenty-four hours you can begin to handle your birds, but they should not be away from the warmth of the heat lamp for more than a few minutes until their feathers have started to come in, which happens in a week or two.

If your chicks arrived by mail, they have likely never had anything to eat or drink in their short lives. They are shipped as day-old chicks and can live on stored energy they obtained from their egg yolk for the duration of a two- to three-day shipping process with few problems. A few mail-order companies do ship their chicks with a gel that provides some food and moisture on their journey.

Once home, chicks will usually find their feeder and water without difficulty and begin to eat and drink almost immediately. However, if you notice a chick that simply does not seem to be picking up the routine, it might help to pick the little gal up and dip her beak in the water. She will act shocked and affronted, but will soon recover, and will likely begin to drink thirstily. Chicks from a feed or specialty store will likely come into your lives knowing the drill, having learned from their brooder-mates or already having been beak-dipped. At the store, we usually take the lid off the feeder and put the chicks directly into it to show them where the food is. You can try this at home for a few hours, but put the lid back on before they make a mess.

THE ART OF CHICK MAINTENANCE

Beyond setting up the brooder, you will need to perform several maintenance tasks on a regular basis, and be on the lookout for a few common health problems. It sounds simple, and it is, but there seem to be as many ways of caring for chicks as there are chicken keepers. We call it an art (not a science) because there is precious little quality agricultural research available that's much help to the backyard chicken keeper. As a result, there are numerous approaches to raising chicks, and much of our advice will contradict what you read elsewhere. What we present here are the methods that have worked best for us, our customers, and the thousands of chicks they have successfully raised. Gather information from as many sources as you like—you'll likely find that no one authority has covered all of the possibilities. You'll decide what works for you, and those will be the only "right" answers.

Food and Water

Besides the snug, safe brooder with its warm and cool zones, you will need to provide food and water for your chicks. At the store we prefer organic feeds for adults, but for chicks we usually recommend that you use a conventional type of feed called "medicated starter." That's because there is a nasty bug called coccidia that is the biggest killer of chicks besides stress (they often go hand in hand), and providing medicated feed is the simplest way to help your chicks avoid this common malady. It has an ingredient that prevents excessive proliferation of the organism (but does not eradicate it), allowing the chicks to slowly develop immunity to the tiny gut-dwelling parasites by about fourteen weeks. As mentioned earlier, there is a vaccine available that eliminates the need for medicated feed, but it is new enough that most hatcheries do not offer it. Assume that you need medicated feed unless you have been assured by your chick supplier that your chicks have been vaccinated.

Many folks are aware of a growing problem called antibiotic resistance, which is caused by routinely dosing livestock with antibiotics to compensate for unhealthy environments. This is different—the medication in this kind of chick feed is not an antibiotic and does not cause this problem. We readily acknowledge that many people we know have had great success using nonmedicated feed. However, we estimate that the use of a medicated feed is responsible for a 15 percent increase in survival. That may not sound like much, but multiply that percentage by the thousands of chicks we sell and you can see why we feel so strongly about recommending the medicated feed.

Whichever type of feed you choose, your birds should remain on chick starter feed until they are about twelve weeks old. After that, they can be moved to "developer" feed, which contains less protein than chick starter and less calcium than adult layer feed. If you can't find developer, you can mix starter feed and layer feed fifty-fifty. Either way, you will switch to layer feed at about twenty weeks. You'll find much more information on feed in chapter 6.

The chicks can start eating treats in addition to their regular feed almost immediately, but be selective and keep this a small part of their diet. Initially their guts are tiny, so they can't digest anything too complex, and the starter feed is what they really need the most of to develop properly. If you want to treat, start with hard-boiled egg, finely chopped. No, that's not cannibalism! To the contrary: eggs are designed for chick nutrition, not just for us to eat. After a couple of weeks they can start eating bits of lettuce or other greens, especially if they get chick grit (described shortly). Small mealworms from pet shops are a good source of protein, and chicks seem to love them. Avoid larger earthworms until the birds have moved outside, as they are a potential source of parasites that your chicks cannot yet tolerate. We also recommend avoiding bread and other processed foods for now. Chick scratch (coarse grains the texture of polenta) is sometimes available and is safe to feed at any age.

Fortunately, water is a much simpler matter. Make sure it's cool, clean, and changed about once a day—and clear debris from it and top it off twice a day. That's it! You may want to use a vitamin and electrolyte powder dissolved into the water (it's like a chicken version of sports rehydration drinks) for the first few weeks to ensure that they are getting everything they need. If you do, be extra vigilant about cleaning the water, as bacteria will grow faster in the enhanced water. To prevent spills, place the water dispenser on a firm patch of shavings (more on this shortly) and make sure that it is level.

As the chicks get bigger, the water can be raised a bit to elevate it above the litter, keeping it cleaner. If using a cardboard box, consider placing a small plate filled with shavings under the fount to keep it from soaking through.

Grit, or small rocks, may be offered to chicks or mixed into their food. Adult birds swallow small stones to help "chew" their food in their gut, and many believe that providing miniature stones to chicks helps their digestion and may help prevent pasting up of the vent (more on this shortly). We provide grit to our chicks either by offering a small dish of grit to be ingested as desired or by mixing chick grit at about a 1:25 ratio into their food (1 pound of grit in 25 pounds of feed).

Sanitation

Despite the best efforts of the good folks who design chick supplies, your chicks will eventually succeed in making a mess of their food and water dispensers, so you will need to check on them at least twice a day. Few things will threaten the health of a baby chick faster than a waterer full of poop. Remove any visible mess from the waterer with your fingers whenever it's convenient each day (this takes only moments if done morning and evening); once a day wash it out more thoroughly and refill with cool, clean water. The feeder needs only to be cleared of debris and refilled every day or two and washed if visibly dirty or between batches of chicks.

When we set up a new brooder, we scatter a layer of pine shavings approximately two inches deep over the brooder floor; we add additional layers approximately every other day, depending on the level of soiling (which relates to the size and number of birds you have), then completely change the litter periodically as described next. This keeps the chicks in a consistently clean and dry environment, and keeps odors practically nonexistent.

About once a week, you should scoop out all the litter from the brooder box and start fresh. Remove any litter that has become wet as soon as you notice it, though, because this promotes mold and odor. If you have chosen to use non-medicated starter feed (all organic choices fall into this category), you will need to be even more vigilant in promoting brooder cleanliness. The task of protecting the chicks from excessive coccidia falls to you alone, and the surest way to do this is to make sure the chicks are never ingesting traces of their own droppings.

When you make any changes to their litter, the chicks will act like the sky is falling, but don't worry—they will forget their trauma within moments. An approach that works well for us is to scoop out half the brooder with a dust pan and lay down clean litter on that side. Then shoo your chicks over to the clean side and scoop out the other half. This gives them an area of refuge throughout the process. We recommend wearing a dust mask for this procedure.

Getting to Know You

Raising chicks is not all about work. After the chicks are in their brooder and have had a day or so to acclimate, it is time to interact! If you handle your chicks regularly, they are much more likely to grow into adult birds who are comfortable with being held. However, baby chicks are somewhat delicate, and there are definitely wrong and right ways to hold them. A good first step is to let the chicks get to know your hand. Lower your hand into the brooder about an inch off the ground, and after the chicks get used to its presence, they will likely be hopping all over it. This is a good approach for children to take, as there is very low risk for chick injury with this approach. When you decide to pick up your chicks, be sure not to squeeze or otherwise apply too much pressure to their little bodies.

When handling chicks, we find that it is helpful to imagine that you are making a loose cage with the fingers of one hand. Your fingers should be close enough together that the chick cannot escape, but not actually putting any pressure on your tiny bird. Support the chick with one hand below and cup one hand above, allowing her cute little head to poke up between your thumb and forefinger. Keep her near your belly for warmth and stability. Adorable and safe! (See photo.) Keep in mind that the higher above the floor you hold the chick, the farther she will fall if the wily little gal wriggles loose.

One of the best tricks we have learned from a chicken aficionado friend of ours is how to correctly pet any chicken or chick. Most chicks and hens will instinctively crouch and tense up if you pet their heads or backs. However, if you gently stroke their chests and under their necks, they seem to relax and even enjoy the attention. Also, this is a particularly soft part of any chicken, so it feels good for both of you.

Chick Behavior

Besides petting and holding them, you will enjoy simply watching your chicks interact with each other and their environment. Most of their time will be spent eating, drinking, scratching in their litter, and sleeping. You will notice that they have quite short sleeping and waking cycles. Often, all your chicks will be sleeping or awake at the same time. This is primarily because any chicks that are awake will indiscriminately walk all over their sleeping cohorts, startling them from sleep. There simply is not much rest for a chick unless the whole brood is sleeping.

Chicks tend to fall asleep quite suddenly, often falling over where they stand after a little swaying. Shortly after we first opened our store, we heard a shriek from the chick area, and rushed to find a horrified little girl pointing to a motionless chick. "It just died!" she exclaimed. Upon further inspection, and a gentle prod, the chick sprang to life with an offended cheep.

Sometimes our store will get a phone call from a distraught novice chick keeper who reports that one of her chicks is bullying the others. Chicks do sometimes become overly fascinated with each other's eyes or toes, and chicks do peck at anything they find potentially interesting. Once we had a chick who seemed convinced that the toes of her brooder-mates were actually worms. She would reach out, grab the victim's toe, and yank with all her might. The object of her fascination would be pulled off her feet and cheep loudly in protest. This went on for days. We had never seen anything like it. We felt terrible for the other chicks, and we tried all sorts of tricks to stop it. We made a little brooder just for the trouble-making chick, but she was terribly lonely and let us know it. We tried putting only one other chick in with her, but you can imagine how the other chick felt about that. We tried separating her from the other chicks with a see-through mesh, but she still felt lonely. Finally, she simply grew out of it, and the other chicks didn't seem to be emotionally traumatized or physically injured. To this day, we call that little trouble-maker "Yank-ee," and she has grown up to be one of our favorite chickens.

In general, unless actual physical harm is being done, try to keep in mind that your chicks are simply behaving as chicks, and their social rules are very different from humans'. On the other hand, we did once take back from a worried customer a chick who was compulsively pecking at the eyes of her cohorts. We put her in a brooder with some slightly older chicks who quickly showed her that such behavior would not be tolerated. She also grew into a delightful hen.

Chick Health

If you follow the recommendations we have just discussed, chances are you will have a worry-free experience as your tiny flock grows. However, you should keep a sharp eye (and ear) out for any possible health problems and take corrective action as soon as possible. Often chicks will alert you to an unhealthy situation before you notice symptoms. Contented chicks make some noise, but it is a mellow, low-toned cheeping that is charming to hear. A chick that is too cold, hot, hungry, thirsty, lonely, or otherwise stressed will cheep in a loud, high-pitched manner that should catch your attention. If you intervene at this stage, you may be able to remedy the situation and prevent illness.

Keep It Clean

It is important to never touch your hands to your face while handling your chicks and always wash your hands afterward. As with any pet, there is a potential for disease transmission (salmonella is the bacteria most associated with chickens), and you should take reasonable precautions. It's also a good idea to wash your hands before interacting with a chick as a courtesy to their immune systems. For these reasons, we always keep a bottle of alcohol-based hand sanitizer near the brooders and strongly suggest that you do so as well—and use it regularly.

The most common symptom of an unhealthy chick is lethargy or generalized sluggishness. Chicks are perky little creatures, and it is not normal for them to stand still for more than a moment or so unless they are about to drop off to sleep. If one of your chicks is consistently moving slowly or seems unresponsive, something may very well be wrong, and you will want to help as soon as possible.

The first thing to consider is the quality of their environment: Is their water, food, and litter being kept fresh? Are you using a recommended substance as your litter? Are your chicks in a warm, relatively draft-free environment? Are they enjoying a reasonably low-stress environment, or is Fluffy the cat stalking around their brooder intimidating them? If any of these factors needs to be changed, do so immediately.

Beyond environmental issues, the most common ailment you might encounter is pasting up, which we alluded to earlier. This happens when the chick's poop sticks to and blocks up her vent (the multipurpose orifice on a chicken's bottom). We're not talking about a little smudge on the downy butt feathers. Pasting up is usually seen as a hard mass, the size of pencil eraser (or larger) and often just as hard. The chick's vent becomes crusted over and blocked by the droppings, which often cling to the surrounding fluff, further cementing it in place. This sounds uncomfortable, and it is!

We've heard of a number of approaches for preventing pasting up, including adding special chick grit to the chick food (we do this at the store) and adding substances to their water such as molasses, milk, and baking soda. Because we have not tried these water additives ourselves, we can make no recommendation either way. The only thing we know for certain is that keeping well-bred, healthy chicks in a clean, low-stress environment is the best prevention.

Removing the poop is the first step in treating the problem and relieving your chick. You'll feel nervous the first time you undertake this task, but take a deep breath: you will do fine. First, dampen a paper towel with warm (not scalding) water and hold it in one hand. Pick up your affected chick with your other hand, and sit her down on the paper towel so that the blockage is in contact with the damp surface. The goal here is to moisten the often dry and crusty poop. You may have to hold her like this for a minute or two. Usually, the chick finds the warmth of your hands and the towel soothing, and after an initial struggle she will begin to snooze. After the droppings have softened, you may be able to simply wipe it away with the wet towel. If the blockage is very large, you will need to carefully cut the glob with small scissors before soaking. If you do this, be very conservative and stay well away from the skin. Repeat these steps daily or more often as needed to keep her vent clear and unobstructed.

A strange but helpful way to know whether a vent is clear and functioning properly is to blow at the chick's bottom and look for the vent to flare or "wink" at you. We recommend that you not let your friends or spouse see you do this. Sometimes after you clear a pasted-up bottom, your chick will evacuate an impressive quantity of retained poop. Don't worry, this is a great sign; just think of the relief your chick is experiencing at that moment!

Based on feedback from our customers and our own experiences, we have found that a pasted vent can be a sign of a more serious intestinal infection. If the pasting up lasts more than a couple of days, or if there are other symptoms—including green, frothy, foul-smelling, or bloody poop—we would suspect an intestinal infection (usually coccidiosis) and recommend that the whole flock be

treated immediately for this dangerous threat. At the store we use a simple and inexpensive antibacterial medication called Sulmet Drinking Water Solution. Add this (or a similar medication in an appropriate dose) in the quantity of 1 half tablespoon per quart for two days. After that, you can halve the concentration for an additional four days. Using an antimicrobial like this at the first signs of infection is critical because these diseases often progress rapidly and are usually fatal. If your beliefs compel you to use alternative treatments, by all means do so. Just be sure to use the strongest tool in your arsenal and act quickly.

There are a host of other mild to serious conditions that you may encounter; we will cover most of them in chapter 7. Whatever happens, it's important to remember that you are doing your best to keep your chickens healthy, and with your help they will survive at a much higher rate than they would in most large-scale operations.

LEAVING THE NEST

You and the chicks will settle into a graceful and enjoyable routine after a few days. You will delight in their antics, fret over minor changes, and bond more than you ever imagined you could with these odd but adorable little creatures. They will develop very rapidly; soon the baby fuzzballs you brought home will be transformed into much larger, partially feathered, awkward-looking teens. Your chicks will need to remain indoors in their brooder for six to eight weeks (from hatching) as they continue to grow and their adult plumage emerges. We have found that the length of the chicks' stay indoors varies depending on four factors: the size of their brooder, your tolerance for chick-dust in the house, the temperature outside, and the state of completion of your coop.

Put simply, if you buy your chicks early or late in the year (February through early April and September through October in most parts of the country), or you're adding the chicks to an existing flock, you should try to wait until they are a full eight weeks old before moving them outside. For chicks hatched mid-April to August, you can probably go with six weeks. If you are starting in late fall, you'll need to provide supplemental heat when moving them outside unless you live in one of the warmest parts of the country. A common way of adding heat is to move the brooder lamp into the coop along with the chicks. If you do, locate it high above them and follow all sensible safety precautions. Some books we have read recommend gradually acclimating your chicks to darkness by turning off the light for a few hours a day before removing them from their brooder. This sounds sensible, but we have not found it necessary, and we suggest you do so only if convenient.

As you would imagine, your climate largely determines when you'll start your chicks and when they can be moved outside. If you live in Hawaii, for instance, your chicks can be started virtually anytime and can go out when only a few weeks old. If you live in a very cold winter climate, like that of the upper Midwest, plan on starting your chicks in spring in order to set them out by midsummer. Here in temperate Portland it is possible, but not ideal, to start chicks year-round and move them outside as late as November or as early as February.

If you are unable to move the birds outside in a timely manner, or if your birds outgrow your brooder quickly, consider enlarging their indoor environment. You can combine several large boxes or find other ways to contain the rowdy teens. But overcrowding can lead to a number of problems, so do your best to avoid it.

TRANSITION TO OUTDOORS

When the happy day comes to put the chickens outside, we recommend confining them to their new coop at first and using it as an outdoor version of the brooder. After a few days, give them access to their outdoor run as well. The chicks are much larger now and able to handle most spring and summer weather conditions, but at this age they are still highly vulnerable to cats, birds of prey, and many other dangers. We recommend that you not allow them free range of the yard except for brief supervised visits until they are beginning to look like

Pullets enjoying their freedom

adult chickens—usually at about twelve weeks. Or as Hannah says, until they can intimidate a cat (and they will!).

If you are introducing new chicks into an existing flock, you should wait until they are large enough to endure and evade the inevitable harassment of their new flock-mates (at least eight to ten weeks). For many of us, the first time we witness this rough treatment it looks rude or even shocking. In fact, it's quite normal: the older and by default higher-ranking chickens must assert their place in the pecking order or risk losing it. If you give your chickens cover, ample room, and feed, this ritual will involve little more than posturing and chasing the little ones away from the food until the adults have eaten their fill. The teens will soon learn to steer clear of the adults, and everything will be fine. They will form their own subflock and gradually assimilate into the group. Occasionally, there are more serious problems that can even result in injury to the new birds. We will offer a few useful suggestions for managing this and go into greater detail on this subject in chapter 6.

5

COOPS AND RUNS: BUILDING YOUR POULTRY PALACE

BECAUSE CHICKENS NEED PROTECTION, both from the elements and from predators, they need to live in a chicken-scaled house and yard. We call the house a "coop" (or henhouse) and the fenced yard a "run." These can be elaborate structures designed to match your home or other theme, or as simple as a shack—just be sure to build them well enough to last and serve their purpose. We have one customer who built his coop with cedar siding, a moat and drawbridge, and radiant floor heating. Others knock together some boards over a weekend to provide perfectly adequate shelter. If you're not handy or have no time, a contractor will happily do all the work for you, or you might choose to buy a prebuilt coop and set it up yourself; these generally take less than an hour to assemble. Truly, there are as many ways of providing homes for chickens as there are chicken keepers.

In this chapter, we share what we have learned about building housing for flocks that will help to keep them safe and healthy. If you prefer to purchase a coop, consult your local feed store or our Resources section for acceptable options, from the mod but pricey Eglu to converted doghouses. We discuss proper placement of the coop and run (hint: not under the neighbor's bedroom window!) and provide several alternatives for materials and construction techniques. We also provide full plans with step-by-step instructions for a simple and practical coop design.

WHY WE COOP THEM UP

The phrase "cooped up" has negative connotations, bringing to mind confined, tight quarters. To a chicken, however, having a dry, draft-free, protected place to sleep at night is very important. Once darkness falls, chickens are entirely helpless, as their metabolism slows down dramatically. In fact, most chickens enter a near comatose state while sleeping and remain this way even when they are awakened briefly at night. This is handy if you have to clip their wings or perform some other task that requires handling your birds. It is not so great for the chicken if she gets stuck outside at night. If it's raining, she'll get soaked to the bone. If it's snowing, she will be covered by a blanket of snow, unable to react to the situation. If there is a raccoon around, she can't run, fly, or put up a fight.

Coops for peace

Luckily, chickens have an instinct that drives them to return to a dark, protected area every night. Your job is to provide it for them and make sure that they are securely inside without fail. Although the instinct to "come home to roost" is strong, some chickens will need a little help. A friend of ours had to pick up each chicken from where they were huddled outside the coop door and place them on the perch for several weeks before they started doing it on their own. This is not typical, however; we think these chickens just liked the attention.

ON THE RUN

In addition to the coop, your flock will need a run. This fenced pen is needed to confine your birds outdoors during some or all of the day. When planning, it is helpful to decide early on how much time your chickens will be allowed to roam the yard freely. We release our hens almost every morning to patrol the backyard during the day. Other chicken keepers, such as people who live among large daytime predator populations or those wanting to reduce the birds' impact on their yard, will keep their birds confined in the run for most or all of their lives. This is an important distinction—how you decide to handle it will affect both the size and security level of what you build.

In our situation, the run is small (4 by 8 feet) and lightly protected by chicken wire. It need not be very large or secure, because the birds are seldom inside it for very long. For us, it's most important to protect our birds from nocturnal predators, so we focused on building a safe, sturdy coop that can be isolated from the less-secure run at night. If we wanted to accommodate our flock of seven hens inside the run exclusively during the day, it would need to be at least twice the size (maybe 6 by 10 feet) and better able to protect the confined and vulnerable birds from attack. The greater size would also be needed to spread their waste and to minimize conflicts within the flock. We offer detailed size and protection guidelines later in this chapter.

As we noted earlier, it's important to consider how your birds will move between the coop and the run. If you enclose the flock completely inside the coop each night (as we suggest), you will control when they can leave the coop and access either the run or the rest of your yard. However, many people leave the door between the coop and the run open at all times, allowing free access between them. If you do this, the run must be as secure as the coop to prevent predators from entering the coop through the run. Although this is a given for those who will build a large, secure run for constant occupancy, those considering building a flimsier run need to keep this relationship in mind and either build a sturdier run than they anticipated or resolve to keep the coop door closed each and every night.

CONSTRUCTION OF COOPS AND RUNS

Whether you do it in style or on the cheap, build your coop and run to be sturdy and serve their functions, and you will thank yourself down the road. Feel free to express yourself—build in a unique shape or use reclaimed lumber, for instance—just build it correctly. What does that mean? First, it must be secure:

An integrated coop and run

there should be no easy entry points for predators in either the coop or the run. Of equal importance in most climates, the coop should be snug and protect your birds from the elements; at least part of the run should be sheltered on three sides from rain or snow, but need not be as weatherproof as the coop. Lastly, use appropriate construction techniques such as joining lumber properly, protecting raw wood, and using treated wood or concrete blocks to keep your structures from rotting where they meet the ground.

Keep in mind that larger coops and runs will often require that you sink posts into the ground with a concrete or gravel base to support them. Although this type of construction is not especially difficult, it will be more like the building of a shed or fence. You can greatly speed up construction time by renting tools for digging and fastening; we think they are worth every penny.

Placement

Deliberate placement of your coop and run is a critically important yet often overlooked decision. As we mentioned in an earlier chapter, the first consideration should be compliance with local rules and regulations. This usually boils down to simply placing the coop and run toward the back of your property or otherwise locating it as far as possible from neighboring dwellings. By keeping the coop a good distance from your neighbors, you will greatly reduce the likelihood of conflicts that can take much of the joy out of chicken keeping. Even if your current neighbors love your chickens, there's no guarantee that their successors will. Locating your coop away from others' homes will also help maintain the peace by keeping any possible odors (there shouldn't be much) and early morning sounds to a minimum. That said, it is important to place the coop and run in an accessible place that you won't mind visiting early in the morning, late at night, and in poor weather conditions.

You should also consider environmental factors such as drainage, sun and shade, and winds. Ideally, your coop and run would sit on moderately sloping, well-drained ground that's not prone to turning to oozing mud when it rains. If a swamp is what you've got to work with, you may wish to build a structure elevated above ground level. Consistently wet conditions are a recipe for infection and unhappy birds. Conversely, dry and dusty soil brings its own set of problems, such as inhalation of harmful particles, so it should also be avoided. Aside from such unusual extremes, almost any spot in a typical backyard is fine for chickens.

Sun and shade are also important considerations; again, moderation is the key. Ideally, you would locate your hens' home in a bright location that offers plenty of morning sun and some afternoon shade. If we had a choice between deep shade and unrelenting sun, we'd go with shade, because excessive heat, like cold and wet conditions, can make your birds sick and unproductive. It's probably telling that more of our customers reported losing chickens during a recent heat wave than in the severe cold snap our area experienced a few months later. Similarly, although some air movement is desirable, you should try to avoid a location that exposes your birds to chilling winter winds. We highly recommend planting vines to climb on the fence surrounding the run; this can help to moderate both sun and wind. Just choose a nontoxic plant, because your birds will surely sample the foliage. Our favorite plant for training onto a run's fencing is the grape. It's adaptable to many climates and provides the double benefits of shade and fruit. As with any new plantings near chickens, the freshly disturbed soil around the vines' roots should always be protected with heavy objects, such as rocks or logs. This is because chickens love to scratch loose soil and will damage the roots if they're not covered. Similarly, the stems should be tied to the fencing and the leaves shielded (with netting or fencing material) from grazing birds for the first 2 or 3 vertical feet. As the vines grow taller this protection can be removed.

Grape vines

Size and Time Commitment

The size of your coop and run depends primarily on the number of birds you will be housing and the amount of time they will spend inside these structures. Rather than provide a chart, we think it's more helpful to offer general guidelines and allow you to make adjustments for your own unique situation. That said, for a backyard flock of three or four standard-sized hens, we recommend a minimum coop size of about 3 feet cubed (3 by 3 by 3 feet, or 27 cubic feet). Ideally, it would be larger to reduce cleaning needs and accommodate future flock expansion. Our flock of seven resides very comfortably in a 4-foot cubed coop, and there's room for a few more. Any more birds than that and we'd need a coop that's large enough to walk into—resembling a shed more than a coop. (Knowing us, this is only a matter of time.)

The size of the run will also depend largely on the size of your flock and the amount of time they will spend in it. For a temporary holding area for three hens, a minimum size is about 2 feet high, 3 feet wide, and 6 feet long. That's pretty snug, but it will do the job if they are in there for only a few hours at a time, such as in the morning before you let them out into the yard to range freely or when they need to be confined while you're entertaining in the backyard (they may steal food!). To accommodate a few more hens on a temporary basis, consider making the run 4 feet high, 4 feet wide, and 8 feet long (this is also conveniently closer to common lumber dimensions).

If you want to have more than a few hens or will need to keep them in the run for long periods of time, we suggest a minimum run size of 6 feet high,

Build it BIG

We urge you to build your coop and run to the largest reasonable dimensions you can manage rather than the smallest you think you can get away with, for two important reasons. First, you are likely going to enjoy this hobby, and once you get the bug you'll want to add new birds to your flock to enjoy new breeds or get more eggs. It is much easier to build a roomier coop and run in the first place than to try to augment it (or worse, realize you need to build a whole new setup). Second, as your flock ages you will want to add a couple of new hens now and again to keep egg productivity steady. Even if you intend to stick with the usual three birds, your older ones will be around for several years overlapping the new arrivals, causing your flock to swell periodically. A spacious, uncrowded coop and run setup will also be more sanitary and require less frequent cleaning.

6 feet wide, and 8 feet long. This added vertical space will allow you easy, walk-in access to the run for cleaning, collecting stray eggs, and catching birds. It will also give you the option to provide perches that expand the usable volume of the area for your hens substantially. We plan to increase our own run size to 10 feet wide and 35 feet long. This will confine our flock to the back half of our yard to keep our patio cleaner.

Beyond the number of birds and the time they will spend inside, the coop and run sizes you choose will vary depending on available time and materials. Building a coop and run of any size is surprisingly time consuming. If you are not a professional builder, plan on spending several weekends constructing these. We emphasize this so that you will get an early start and will be in a position to move your young birds outside as soon as they are ready. Whatever size you choose, we recommend that you base your dimensions on the sizes of commonly available lumber to minimize waste and cost.

THE ESSENTIAL ELEMENTS

It's helpful to think of your coop and run requirements in the context of four elements: shelter, protection, daily needs, and maintenance. In this section we examine these elements and see how they can work together.

Shelter

In cool climates, coops act as a sanctuary for your birds, where they can safely sleep, preen, and lay their eggs while minimizing the energy drag of keeping warm. For this reason, coops should be built with four walls, a floor, and a roof. That may seem obvious, but we have had a small number of customers tell us that they are using coops with only three sides with a fourth side made of mesh. Still others have coops with an open floor or ceiling. Although your birds may survive in such conditions, they are unlikely to truly thrive in such a coop when nasty weather arrives. Rain and wind will find these openings, leaving your birds little

Features of a Well-Built Coop and Run

Coop:

- Rain-tight and waterproofed
- Draft-free (but not necessarily insulated)
- Well ventilated (near top, not mid-side)
- Perch (1 to 2 inches wide and 2 to 4 feet long)
- Nesting box (one for every three birds)
- Easy to clean (easy access)
- Door(s) with predator-resistant latch(es)

Run:

- Sturdy, protective fencing with small openings (stronger and tighter if it's your flock's only outdoor environment)
- Protection from rain (including drier area for feeding)
- Door with predator-resistant latch
- Height sufficient for access by the hens, and by the chicken keeper if so desired

better off than they would be outdoors. However, in very warm climates a more open structure may work better than in more temperate or cold regions.

At the other extreme, we have seen some eager coop makers overbuild these simple structures to the point that they might do more harm than good. For instance, we have talked to some folks who have built their coop to be so airtight that moisture and gases from the birds and their droppings can't escape. Other chicken keepers, in their zeal to make the coop resemble a human home, add so many windows that the coop acts as a greenhouse and traps too much heat.

It is preferable to find a balance between keeping out the elements and sealing in the birds completely. Although each coop is different, we have found the following approaches to weatherproofing to be helpful:

1. Build your coop using only solid plywood; minimize gaps, but provide little or no additional insulation.
2. Prime, paint, or otherwise protect all exterior walls.
3. For flooring, use a water-resistant material such as sheet linoleum or vinyl inside the coop, or seal with nontoxic wood preservative.

4. For the roof, use composite, plastic, or corrugated metal roofing or another solid, weatherproof material such as primed and painted plywood.
5. Provide 1–2 inches of ventilation near the top of the coop walls under the roof overhang.

Venting near the top of the walls under the roof overhang is an important feature that allows moisture and gases to escape. This vent location will act like a chimney while leaving most of the solid walls to protect the sleeping birds on their roosts from wind and blowing rain. Some folks also like to incorporate windows for ventilation and viewing purposes (a real peep show!). This can work; just be sure that you consider how the coop can be ventilated when the windows need to be closed, such as on a cold winter day. As discussed next, vents and windows will need to be screened with hardware cloth to prevent access by predators.

If the coop meets these criteria, you can feel free to experiment with design and materials. We have seen some coops with walls made of corrugated metal, others of rough-hewn cedar planks, and even one built from a repurposed recycling bin; use whatever appropriate materials you have on hand. The coop can, and probably should, reflect your regional building styles. In the Southwest, an adobe coop with a tile roof might be just the ticket to keep temperatures moderate inside the coop. In New England, a whitewashed saltbox coop would be attractive and functional and also blend neatly into its neighborhood context.

If you live in an area of high rainfall like Portland, you will want to slope the roof moderately and consider using a gutter to divert the deluge away from the area around the coop. To reduce trudging through mud, it's wise to slope your roof away from the side from which you will access the coop. In snowy areas, consider an A-frame, which will naturally shed snow loads that could collapse the structure. Again, take your cues from your regional building styles.

If you live in a very cold climate, you would be wise to make provision for supplemental heat and possibly light (see chapter 8) while designing your coop. This may mean simply providing a place to clip a chick-warming lamp near the roosts, or it could be as elaborate as the setup established by one of our customers. He made a radiant floor heating system for his coop by creating a subfloor heated by two 40-watt lightbulbs. To power any artificial heat source, you will need access to an electrical outlet. If you are building an elaborate coop, go ahead and wire in an electrical outlet if you wish. Otherwise, it is a good idea to build your coop within reach of an outdoor outlet. We have also seen some folks use solar energy systems; if your coop location gets enough sun in the colder months, we encourage you to try one.

The run need not be nearly as sheltered as the coop. At a minimum, you should provide a roof over some or all of the run's length, again sloping it away from the side from which you'll access it. If possible, create an area of the run adjacent to the coop that is sheltered on three sides from wind and windblown rain while still being accessible. We like to call this area the "porch." Your chickens will like eating on the porch, as we will describe in the "Daily Needs" section (see page 83). If nothing else, use a tarp to cover part of the run seasonally.

Protection

In every neighborhood there are one or more resident predators that would be thrilled to eat or otherwise attack your chickens. It's your job to stop them. To do this, you need to provide protective structures that keep predators at bay when you are not around. We like to think of our protective system as rings of defense.

Fenced Yard

The outermost ring of defense is the fence around your yard. Although it's possible to keep chickens in an unfenced yard, we strongly advise against it. A typical six-foot-high wooden fence in good repair will do an admirable job of protecting your yard from one of the most feared daytime predators of chickens: loose domestic dogs. Sadly, we have lost more chickens to unleashed dogs than to any other predator. This has led to periodic improvements, such as raising the height of our gates to a full six feet and adding

Hardware cloth fencing

hardware cloth to the base of the fence adjoining a park so dogs are unable to dig under it. A fence of this height will also keep most standard-breed hens from flying out (see "Flights of Fancy," page 117).

Run

The run is the next line of defense. How secure you should make it depends on several related factors. First, assess the likely level of threat to your birds. In any situation, there are at least some animals that pose a threat to your chickens. If you live near a park or open space or in a semirural area, there are likely even more wily creatures looking to make a quick meal of your birds. In these situations, it is advisable to keep your chickens in a protective structure both day and night. If you live in a more urban setting, it is generally safe to let your birds wander your yard in the daytime to forage or remain in a lightly fortified run.

If you decide that it's safe to allow your birds to roam the yard, your run will be a temporary or seldom-used holding area for them and will not need to be as carefully constructed. If they will need to spend most or all of their lives inside the coop and run, you will want to make them both very secure. If you're not sure, it's better to build a run that can withstand attacks from determined predators.

Building a secure run means using strong materials in a way that leaves no easy means of entry once the door has been closed. Chicken wire, the classic run building material, is really better suited for keeping chickens inside than it is for keeping predators out. We have listened to several firsthand accounts of raccoons

Carport Frames as Runs

Our clever friends Chris and Tonya repurposed a used single-car portable carport frame as a run. At about 10 by 20 feet, it is the perfect size for a dozen chickens or more and can easily be moved if needed. They simply covered the frame in chicken wire on the sides and top, with hardware cloth for the bottom 3 feet. If you try this, we recommend that you follow the other design requirements mentioned in this chapter, such as covering it in winter with a large tarp for shelter from the elements.

Buried fencing method

reaching through the relatively large openings in chicken wire to snag chickens inside. For this reason, we recommend it only for runs that need to be lightly secured. The fencing we do recommend is hardware cloth, which is available in a half-inch size that is both strong and easy to work with. Although it's relatively expensive, we strongly urge you to use it in any situation where security is of primary concern. At a minimum, you can use this material around the bottom 3 feet of the run to provide extra protection where it's needed most. The wooden frame for the run should also be sturdy and provide a good base on which to stretch the fencing. If you hold a corner and push, there should be very little give.

If you are building a high-security run, you will also need to prevent animals from digging under it. There are at least three approaches:

1. Bury the bottom 12 to 18 inches of perimeter fencing in a ditch and backfill. This economical, though laborious, method provides good predator protection but may not be practical if your soil is hard to dig.
2. Line the bottom of the run with hardware cloth and bury it in gravel. This is the method we used on our run, and it's worked well for us. It avoids the labor of digging and backfilling a trench, but it increases the cost of materials.

3. Extend the fencing to make a "skirt" around the whole base of the run and cover it with heavy wood or rocks. This combines the best of the other two methods and is now our top recommendation.

Coop

The innermost line of defense—the sanctuary—is the coop itself. While sleeping in the coop, your birds will rely completely on the protection it provides. If you have built a sturdy coop and incorporated the features designed to keep out the elements, you are on your way to having a secure coop as well.

To harm your birds, predators will first need to reach them. If the walls are made of sturdy materials that are in good repair, the only entry points will be through the vents, windows, and doors. Securing the small vents is a simple matter of covering them with hardware cloth from the interior of the coop where they cannot be easily ripped free by a probing claw.

Spring-loaded latch closure

Windows and doors need to be secured with latches or locked clasps. We have found the most useful type of latch to be the spring-loaded hook-and-eye design. This inexpensive hardware is not only easy to secure but also easy to unlatch on a door that must be opened daily. This kind of latch may be impractical in cold regions if the tiny metal parts regularly get covered in ice, or in areas near the sea where metal is prone to rusting—again, use what works in your area. Clasps and locks have the advantage of being simpler mechanically and very secure, but locking and unlocking frequently used doors can be annoying. Whatever you choose, select something that is unlikely to opened by a lucky swipe of a paw or snout. As a general rule, use something you need thumbs to operate! Finally, a motion-sensing outdoor light near the coop may help deter nocturnal predators.

Daily Needs

Your coop and run will need several features to provide for your chickens' daily functions. When these are in place and working properly, the flock will cluck cheerfully along, laying well and in good health. If one or more of these elements is missing or in poor repair, trouble usually follows.

Coop Amenities

In addition to protection and shelter, your coop will house a few important items that your chickens will use daily.

ROOST First and foremost, you must provide a roost. This feature is simply a perch for your birds to rest and sleep on; essentially, it is a man-made tree branch. Roosts are easy to provide by securing a 2-inch closet dowel horizontally 1 to 3 feet above the floor of your coop. Some coop builders prefer to use flat or slightly rounded-edge boards for roosting. The theory is that a flatter roost allows the birds' feet to be flat instead of curled around a round perch. This may allow them to more effectively protect their feet with their warm feathers in cold weather. However, the edges must not be sharp, or they could cause injury to your birds' feet. We have always used round roosts and had no trouble. Whichever type of roost you choose, provide 8 to 12 inches of roost length per bird.

Because your birds will do much of their pooping at night while roosting, some people add a droppings board or box under the roost to collect the mess. This allows you to remove most coop droppings with little effort. By removing this most concentrated accumulation of poop, you can substantially increase the time between general coop cleanings.

NESTING BOXES The other main element of a coop is the nesting box. Hens need these snug, dark places to lay their eggs and have a little privacy from you and the rest of the flock. Premade boxes are available, but most people repurpose cabinets, wood crates, or cubbyholes. Nests can be located inside the coop or can be attached to the outside, usually positioned for easy access without having to enter the run. Either way, they should be about 12 inches high, 12 inches wide, and 12 inches deep. A small lip is useful to keep eggs from rolling out and

Coop interiors with a hen leaving the perch (left) and two nesting boxes (right)

to hold nesting materials in place (we suggest that you use pine shavings or hay, not straw). When designing your coop, try to place the boxes in a location where you will have easy access for egg collecting. The nest boxes must also be protected from rain and snow. We suggest that you provide one nest box for every three hens. Although it seems logical to provide one for each bird, it is unnecessary, because small flocks of chickens seem to prefer laying communally. For a flock of three hens, a second box, while not required, will be used from time to time if provided.

Run

The run is where your flock will be doing the majority of its eating, scratching, and drinking. We do not recommend placing either the feeder or the fount (which provides water) inside the coop unless (1) the coop is large enough to position them well away from the dirtiest (poopiest) areas or (2) you are very fastidious about cleaning regularly. By separating feeders and founts from the most soiled areas, you minimize the potential for cross-contamination, a common cause of many poultry diseases.

WATER FOUNT Juvenile and full-sized chickens need much larger and sturdier vessels for their water than they did when they were fluffy peeps. Poultry keepers refer to them as "founts"—or at any rate the catalogs do. We also call them "waterers," though it sets off our spell checker annoyingly. A fount is simply a container designed to hold water and keep it relatively free of debris while allowing the birds to drink from it. To accomplish this, they are designed with a covered reservoir and a shallow drinking rim—essentially like the chicks' watering gear but larger. Traditionally, they were made from ceramic or tin, but today they are more likely to be galvanized steel or plastic. We much prefer the metal ones, as we have found the plastic ones to be of generally poor design and construction. Expect to pay between $20 and $40 for a good metal fount.

The water can be placed in almost any location inside the run that is not especially well traveled or dirty. The fount should be set on a level concrete block or several bricks to raise it 6 to 8 inches above the floor of the run. Raising the fount will help to reduce the accumulation of debris from your birds' scratching and foraging nearby. We do not recommend hanging your fount if it was not designed for this purpose, as this may damage it. It is okay for the fount to be exposed to rain, of course, so it can be located in an exposed area of the run.

FEEDER As is true of the fount, young hens and adults will need a feeder that is both higher capacity and harder to tip over than the one they used in the brooder. There are two main types: hanging and range feeders. The hanging ones resemble a hanging metal cylinder on top of a pie plate (skip the plastic versions if you want one that will last), usually holding up to 12 pounds of feed. The range feeders look like a tiny metal horse trough—with the unusual addition of a rotating bar to keep the birds from scratching out the feed with their feet—and have a capacity of about 6 pounds. For most folks we recommend the hanging feeder because of its compact design. If you have room in the dry part of your run, or more than three or four birds, we would steer you toward a range feeder, because its length allows many birds to eat simultaneously. We also appreciate that the feed in the range feeder is always easy to access, unlike in the hanging types that rely on food moving from the cylinder to the tray below by the force of gravity—or more often, requiring a vigorous shaking from you each morning as you stand there in your robe and slippers with hungry hens clamoring at your feet.

Unlike placing the water fount, for the feeder you must find a dry, protected location. This is another good reason to make one area of the run sheltered on three sides. An ideal location is under an elevated coop. If this is not practical or your coop is not elevated, you can create a sheltered area elsewhere using any durable building material. Using a tarp seasonally is another common approach to creating a dry area for the feeder.

Like the water fount, we recommend elevating the feeder 6 to 8 inches to keep debris from accumulating in it. Unlike the fount, there are several good feeders available that can be hung from the run supports. Many popular feeders can be used with a lid to prevent birds from climbing in that works only when hanging. Poultry-sized troughs are also available and can be helpful in certain situations, but they cannot be hung.

Maintenance

To thrive, your birds will need to be kept with a reasonable degree of sanitation. To accomplish this, you will need to maintain your birds' environment by periodically cleaning it. Proper design and construction of the coop and run will make this aspect of chicken care much easier.

Coop

The first consideration when designing your coop for easy maintenance is access. Coops are cleaned anywhere from every week to every three months, but they must be cleaned periodically. Build the coop with large enough doors so you can reach the farthest corners to remove soiled litter, dust to control mites, reach wayward eggs, and so on. As mentioned earlier, many find it helpful to use a droppings board or other removable flat surface to collect poop from the area of highest concentration under the sleeping roosts. This can be anything from a piece of plywood to a cookie sheet, lightly covered in litter to prevent sticking, that can be removed frequently and dumped. If you wish, the droppings board can be built into the coop as a removable floor, with or without a wire floor above (not our favorite style, because instead of dropping through the wire to be collected, poop can instead become stuck). Some clever builders have replaced the traditional roosting bars with a roosting box that provides both a roost and a collection area in one unit. For similar reasons, many people opt to use vinyl flooring or other durable material on the floor of the coop. This prevents moisture accumulation from rotting the floor of the coop and also makes cleaning easier.

It's also a good idea to make all components inside the coop removable. If there is a nest box inside the coop, make it self-contained so you can pull it out of the way when cleaning. The same goes for the roosting bar and anything else that might get in the way of your cleaning.

Run

The run is cleaned far less often than the coop, perhaps every three to six months in most cases. Again, access is key. If possible, make the roof of the run high enough to allow you to walk around inside. This way you can stand while mucking out dirty litter or other accumulations. If this is not practical, or if the run is low and scaled for temporary use, consider making the whole structure removable so you can detach and set it aside, gaining access to the area for cleaning.

Another major design feature of a run, besides its protective sides and sheltering roof, is the floor or ground covering. In our coop we have found pea gravel to be a great base for the light use our run receives. We need only to rake away clumps of poop occasionally; most of the rest seems to dry and sink deep among the rocks. The open spaces between the pebbles seems to keep things aerated, greatly reducing odors.

If your run gets more use and gets dirtier than ours, consider using straw (but only in the run, not in the coop). Straw is useful as a run litter because it's cheap, absorbs poop and odors well, and is easy to compost. You can combine a gravel base with straw as a second layer to get the benefits of both.

If you build your run on a hard surface like concrete or pavers, you can simply hose it off when cleaning. We still recommend using a litter or other material on top, as your birds will need to perform natural behaviors, like scratching at the ground, that are possible only on soft surfaces or in litter.

COOP PLANS AND INSTRUCTIONS

This plan for a coop is simple to build and lends itself easily to customization. We have attempted to strike a balance between ease of construction and efficient use of materials. If you're unfamiliar with any of the supplies, tools, or techniques we mention here, virtually any hardware store can offer helpful advice and guidance. For folks better acquainted with home construction techniques, coop construction should be a very straightforward matter. As we get feedback from more people building this coop, we will post updates and revisions to our book's website, www.achickenineveryyard.com. And, of course, you can also post your own feedback there.

TOOLS

Drill (with assorted bits and drivers)
Jigsaw
Pencil
Speed square
Circular or chop saw
Paint brush

MATERIALS LIST

Eight 2" x 2" x 8' untreated wood boards
Three ½" x 4' x 8' plywood sheets (of the best quality you can afford)
One box 1¾" outdoor wood screws
One small roll (36" x 5') ½" hardware cloth
One box construction staples or finish nails
Four 2" cabinet hinges (galvanized and including screws)
Two 2" eye-hooks and latches (spring-loaded type)
One 3' standard-size wood closed dowel with sockets
One pint each nontoxic wood primer and paint (or other nontoxic wood preservatives)
About 4' x 4' worth of roof shingles or other waterproof material scrap (optional)
One 3' x 3' vinyl flooring scrap (optional)

Step 1—Cuts

2" x 2" x 8' boards:

1. Cut four into these segments: 3', 3', 1'10.5"

2. Cut two into these segments: 2'8", 2'8", 2'2.5"

3. Cut two into these segments: 2'11.5", 2'11.5"

Tip: Save small remnant pieces for handles and other minor features.

½" x 4' x 8' plywood:

Cut as shown on plywood cutting diagram (pages 91–93), including interior cuts.

Tip: Use a ½" hole bit to drill corners of the interior cuts, then finish cuts with the jigsaw. Save large cutouts for doors. Save other scraps for making nest boxes and other small features.

Step 2—Assembly

1. Attach 2 x 2 board segments to side pieces as shown (page 94). For these and other attachments, predrill holes for screws and drill from the outside panels toward the inside of the coop (screw heads on plywood side).

2. Attach remaining 2 x 2 board segments to the floor piece as shown.

3. Place one side piece on the ground with 2 x 2 boards facing up. Slide the floor piece into the groove between boards (perpendicular to the side piece) with 2 x 2 boards facing down (away from the slanted top end of the side piece). Secure the pieces together using diagonal screw angles.

4. Turn the assembly onto its side and slide the second side piece into place as shown (page 94). Secure again with diagonal screws *and* from outside of both side panels into the floor sides.

5. Trim hardware cloth to fit 2" vent openings on the front and back panels. Using construction staples or finish nails, secure to the back of these panels.

6. Attach the front and back panels to the rest of the assembled pieces, again predrilling holes and screwing from the plywood sides toward the board sides.

7. Stand up the assembled coop and attach the bottom shelves (important for stability).

8. Attach the front door with two hinges along the top and the side door with two hinges along the bottom, using the screws included with the hinges. Add eye-hooks and latches to secure the doors (these will also serve as handles).

9. Attach dowel sockets about 10" from the coop floor and 10" from the back (lower side) of the coop using their included screws. Cut dowel to interior width of coop and insert into sockets.

10. Attach the roof panel by screwing from above. Use scrap pieces of wood attached to the inside top sides of the side panels if you need more material to screw into.

Step 3—Finishing

1. Prime and paint or stain all exterior wood surfaces (adding any optional trim first). We recommend that you also attach your choice of shingles or other waterproofing material to the roof at this time.

2. Cut and glue down sheet vinyl scrap to fit the interior floor surface.

3. Place the finished coop on blocks, pavers, or gravel to prevent contact with soil and avoid rot.

4. Add nest boxes (not shown on plan—you can use pre-built ones or make them yourself).

Building a run is less easily described, because of the unique circumstances of each yard and the variety of approaches to security that we described earlier in the chapter. See those sections, pages 72–82, for suggestions to get you started. The most important thing to keep in mind is that if the run is a major part of your predator defenses, it should be built to be sturdy enough to resist being breached from the sides, above, and below by digging animals. Set your posts deeply in gravel or concrete and use metal fencing with a fine (1 inch or smaller) mesh buried or flaring to the sides to ensure maximum protection. If your run is only a temporary holding area or if you are blessed to have few roaming predators, you can use a lighter design and focus more on aesthetics or cost. Either way, don't neglect details like doors and latches—they must be easy to use and secure enough to resist a dexterous, furry prowler.

Situating, designing, and building your coop and run correctly from the outset will pay dividends for the life of your flock. It's definitely worth spending the extra time and money to do it right. Always keep security and shelter paramount, but do not neglect maintenance when planning your design, and your birds will never resent being all cooped up!

Cutting Diagram

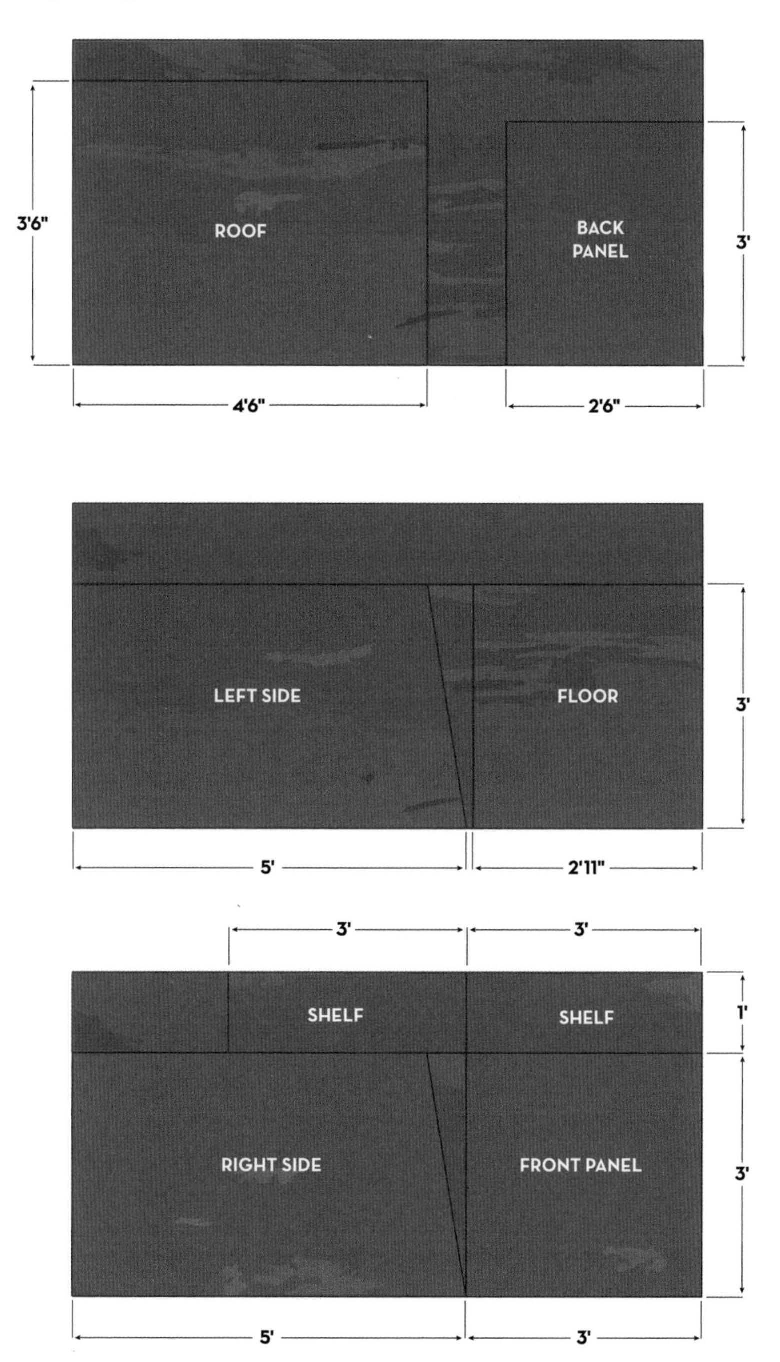

Dimensions

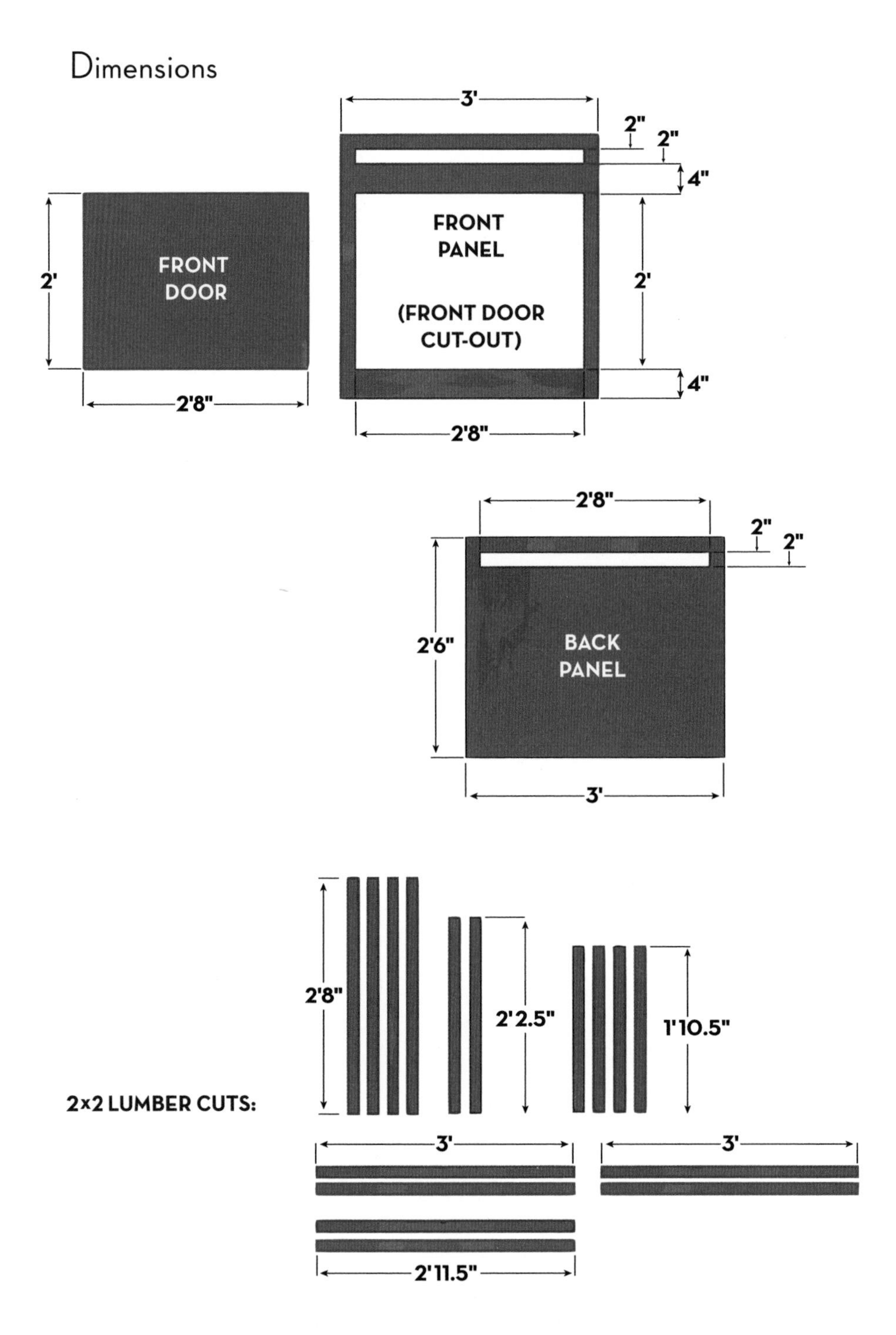
3'
2"
2"
4"
FRONT DOOR
2'
2'8"
FRONT PANEL
(FRONT DOOR CUT-OUT)
2'
4"
2'8"
2'8"
2"
2"
2'6"
BACK PANEL
3'
2x2 LUMBER CUTS:
2'8"
2'2.5"
1'10.5"
3'
3'
2'11.5"

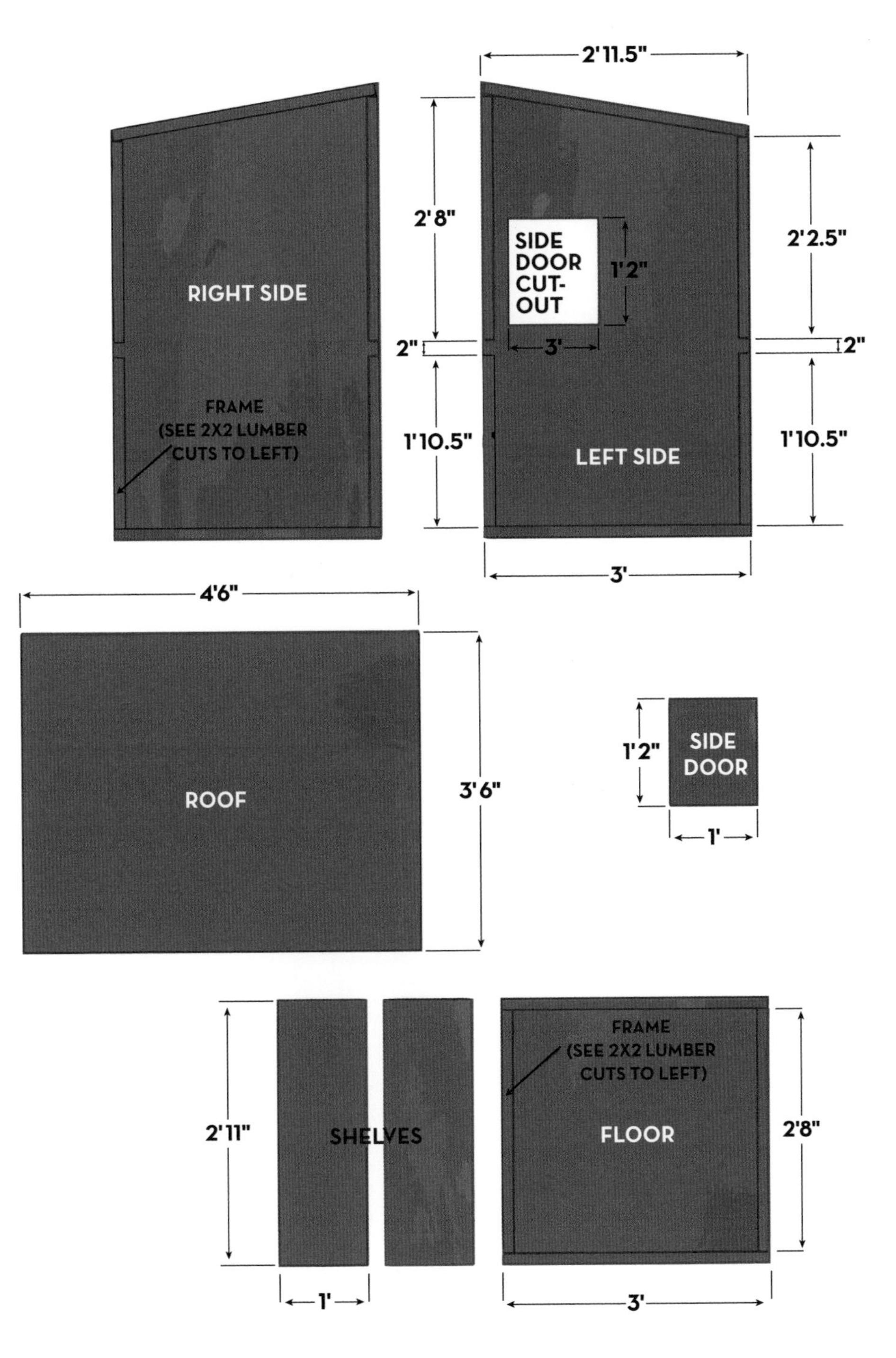
2'11.5"
2'8"
SIDE DOOR CUT-OUT
1'2"
3'
2'2.5"
RIGHT SIDE
2"
2"
FRAME (SEE 2X2 LUMBER CUTS TO LEFT)
1'10.5"
LEFT SIDE
1'10.5"
3'
4'6"
ROOF
3'6"
1'2"
SIDE DOOR
1'
2'11"
SHELVES
FRAME (SEE 2X2 LUMBER CUTS TO LEFT)
FLOOR
2'8"
1'
3'

Assembly

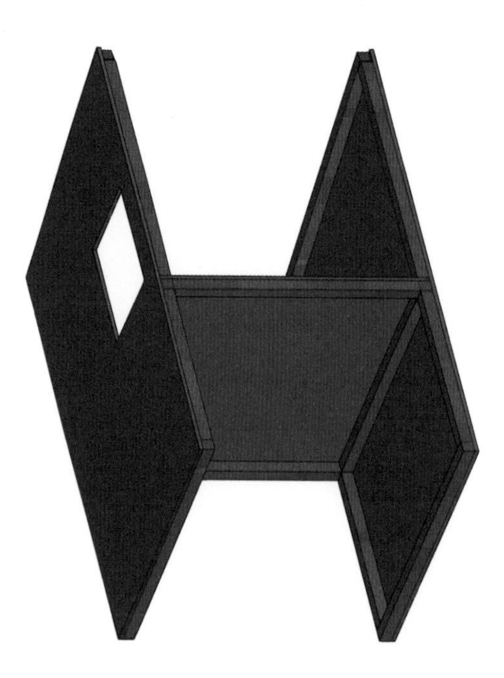

The finished coop (shown with optional trim not in plan)

6

ADULT CHICKENS: GETTING IN TOUCH WITH YOUR INNER FARMER

THE ADORABLE CHICKS you brought home have rapidly lost their fuzz, grown little feathers, and entered a prolonged "awkward teenager" phase. They've gone from falling over to flying over their brooder walls.

Now that the coop and run are completed (right?), it's time to move your young pullets out of their cramped indoor quarters and into the yard where they will have more room. Here they will continue to develop their bodies, tap into their instincts, and with some help from you, stay protected, clean, and well fed.

Just when you've forgotten that there's more to this peculiar pet, they will stop cheeping, start clucking, and then, as if by magic, produce their first eggs. These early eggs may be small, misshapen, or even shell-less, but it will still be a momentous occasion for both you and your newly mature birds. They will seem slightly surprised by this turn of events, and you will undoubtedly be delighted with the prospect of many delicious omelets to come.

Your mature hens will certainly be lower maintenance than they were as baby chicks, but they will have their own unique and equally important requirements. In this chapter we examine the daily routine of chicken keeping, how to anticipate and avoid problems, and the myriad delights of living with your hens.

This transition from chicks to laying hens generally takes four to nine months, depending on breed and season. Hens have needs similar to those of chicks, but they are less tender and better able to fend for themselves. Still, they need good food, fresh water, a sturdy and sanitary shelter, and protection from predators. Provide for these needs, and chicken maintenance will be a mellow and rewarding experience.

ROUTINE

Chicken keeping is not hard to do—there are just a few simple tasks that you must do on a regular basis to keep things humming along nicely. We spend only about ten minutes a day on chicken chores, mostly feeding and collecting eggs. There are other things that need to be done on a less frequent basis. The following summary of activities and their duration and frequency will give you a good idea of what's involved (details later in this chapter). Based on your own unique circumstances, you will develop a routine with your flock that fits your lifestyle.

Here's what we do:

Morning (Five Minutes or Less)

1. Let chickens out of coop and/or run.
2. Check that food and water are available. Top off or refill them and clean if needed.
3. Collect eggs (leaving one in the nest—more on this later).

Evening (Five Minutes or Less)

1. Lock up coop and run after chickens have returned of their own accord (or collect them, if they haven't).
2. Check whether any more eggs have appeared in the nesting box and remove (again leaving one).

Here are the other things that need to be done less frequently:

Weekly

1. Add a layer of shavings to refresh the litter in a process known as the "deep litter" method (two minutes) or, if you prefer to keep things extra tidy, remove and replace all the litter in a process called the "fresh litter" method (ten minutes) (we'll discuss these options in detail later in this chapter).
2. Inspect the coop and run and make repairs as needed (two to thirty minutes).

Hannah cleaning a coop at the store

Monthly

1. Inspect the chickens for signs of illness or pests. Treat or dust if needed (five to thirty minutes).

Every Two to Three Months

1. Remove and replace deep litter (ten to twenty minutes).

Every Four to Six Months

1. Do a thorough cleaning of the coop (one hour).
2. Do a rake-cleaning of the run (ten minutes).

PROTECTING THE REST OF THE YARD

Beyond this low-maintenance chicken haven, we wanted our backyard "farm" to consist of more than chickens. We are also seriously into gardening and growing our own vegetables, so we had to put some effort into integrating these two passions in a small area.

Everything you've heard is true: chickens will descend on your tender garden like locusts! Well, that's a bit of an exaggeration, but they will gladly munch tender perennials and nibble at most veggies if not excluded from that part of your garden. One solution is to keep the chickens in their enclosed run at all times. That's effective, but we liked what the chickens could do for our garden (fertilizing, weeding, bug control, and so on) when they were allowed controlled access to it. For us this meant building a fence around the veggies and admitting the chickens only seasonally. We started with a 2-foot fence (we were so naïve!) but soon graduated to a 4-foot fence, which is enough for most chickens—but not ours. Our highfliers needed a 6-foot wire fence to thwart their vegetable thievery. It actually looks great and makes a useful trellis, so we don't mind.

What we're getting at is that your decisions about how and where to keep your chickens should be dictated by your lifestyle (and in part by your chickens' abilities). If you have a beautifully manicured lawn, an award-winning perennial garden, or stunning statuary out back, you probably do not want your chickens running amok. For you, chickens will be best confined to their equally picturesque coop and run in a tidy corner of your garden. Similarly, if you are a very busy person and frequently away from home, you'll want to set things up to be secure and low-maintenance enough to leave the flock unattended for a couple of days. Alternatively, if you are the kind of person who loves to take care of animals and dote over them, you may opt for a very hands-on approach. If you have a fully fenced yard and you enjoy watching your flock wander about free-range, then simply let them out of their run every day. This is our approach, and it works for us because we can tolerate a little poop on our patio, our vegetable garden is fully fenced, and we think true backyard free-range eggs taste better than any others out there. Simply put, the goal is to develop a strategy for chicken keeping that fits your life and keeps the experience as fun as possible.

Let's now take a closer look at the basic elements of a chicken care system to better inform your personal strategy.

FEEDING—WHAT'S FOR BREAKFAST?

A chicken's life revolves around food. Whether your birds are scratching in the dirt looking for tiny insects or nibbling on grass tips, they will spend nearly every waking moment searching for tasty tidbits. This behavior is all well and good, but it will never provide the full complement of nutrition that laying hens require to produce large numbers of eggs regularly. To get ample eggs and keep your hens in top condition, you will need to provide the bulk of their nutritional needs by serving them commercial feed. For all its familiarity, manufactured feed is a relatively new concept. It's likely that early domestic poultry relied on eating unharvested or spilled grains and legumes, scavenging garbage piles, and grazing pasture for hundreds of years of their history. Feeding was easy, but egg and meat yields were low. As breeders gradually developed hens that could produce more eggs and meat, there arose a need (and opportunity from higher crop yields) to feed them a richer diet that would support their increased metabolic activity. Specialty poultry feeds were born.

Eating from a hanging feeder

Today, commercial farmers rely almost entirely on manufactured feeds for their enormous flocks. Chickens have become an ideal "protein factory" for converting enormous harvests of corn and soy into something more popular to eat. Unlike cattle, who are equipped naturally to graze pasture only, poultry can manage pretty well on a diet dominated by grains and legumes (especially if these are cooked or certain digestive enzymes are added). The main thing critically lacking in this diet, however, are the greens. When hens are removed from the pasture and fed seed crops exclusively, the nutritional quality of both eggs and meat decline substantially. In particular, this results in an undesirable change in the ratio of the omega-3 fats (like those found in fish oil) to the omega-6s (already overabundant in our diets). This goes a long way toward explaining how the egg got its reputation of being full of fats and bad for you. A healthy compromise for both home flocks and small farms is to provide a high-quality feed along with regular access to a lawn or other greens for grazing. Confined birds will enthusiastically show their appreciation for having some greens tossed to them, clucking

Switching Feeds for Adults

In addition to changing what you feed based on life stages, you may occasionally need or want to change the brands or types of feed you give your adult birds. Chickens are creatures of routine, so switching feeds can be disruptive. But it can be done. First, we suggest not blending the two types of feed; we have found that they will usually ignore the new stuff and pick out the familiar type. It's better to just wait until the old feed is gone and change over then. Next, withhold treats for a couple of days to allow your hens to focus on the new menu item. Lastly, it's very important to be patient. Though your birds may turn up their beaks initially, they will eventually eat what you give them, usually within twenty-four hours. We had one customer who complained that her girls were not eating a new feed she had purchased from us. When I asked how long it had been since they ate anything, she replied that they refused to eat the first morning she put it out, so she started to prepare special human food meals for them twice a day. Needless to say, they loved that and refused mere chicken feed. I advised that she stop cooking and wait them out. Now they love the new food and are making do without the home cooking (except on their birthdays!).

with delight as they eagerly shred and devour them. This green material will give a boost to the healthier fats in both the chicken and her eggs. We discuss all of this further in chapter 8.

Above all else, it's important that you provide the appropriate feed for each stage of development. We've already discussed the high-protein, low-calcium starter feed that chicks need until they're about twelve weeks old, and the developer feed, which provides an intermediate balance between starter feed and the layer feed that adults get. Developer is lower in protein than starter, to put the brakes on the chicks' rampant growth, and it's also unmedicated, because by this point the chicks should have developed immunity to the coccidia. Developer differs from the layer feed in being lower in calcium—a mineral that is required in large amounts by laying hens for egg production but is potentially toxic to immature hens who don't yet need much. Developer can sometimes be difficult to find, so, as we mentioned in chapter 4, if you need to improvise, a fifty-fifty mix of unmedicated starter and layer feeds should be just fine. As laying approaches, usually at five to six months, you can gradually increase the proportion of layer in your ration to 100 percent.

FUN FACT

It is fairly common to find double-yolked eggs, but eggs have been found with as many as nine yolks.

Once they are laying, chickens need a feed that is suitable for the demands of near daily egg production. This includes a very specific formula for protein, carbohydrate, and fat along with vitamins and minerals. Most people have little trouble finding conventionally grown layer feed, usually in a pelleted form, from local sources. Depending on where you live, there may be other alternatives available, such as organic or regional blends of available ingredients. Whichever you choose, your layer feed can come in one of three main forms: pellets, crumbles, or mash. Mash consists of a blend of ingredients that have simply been ground and mixed (and often mixed with water prior to feeding to form a moist feed). Pellets are nothing more than mash that has been compressed for easier feeding, and crumbles are just smaller, broken pellets. We recommend that you focus more on the quality of the ingredients and the completeness of the formula than

Roxy and Rosey scratching in our compost; Tweedy grazing our lawn

on the form of the feed. Others may disagree, each stating a strong preference for one type or the other. We suggest that you experiment with several feeds until you find the one that both you and your birds can agree on.

The desire to find something even better to feed our own chickens was the spark that led to our opening the Urban Farm Store. We wanted to give our flock a feed of higher-quality ingredients than large-scale farmers could feed their flocks. And we wanted it to be as local as possible, to conserve resources and support local farmers. We started with a local mill and produced a custom-milled, nonorganic layer feed that contained a large proportion of local ingredients. Eventually, we were able to work with a farmer to formulate a mash-style starter, developer, and layer that were mostly grown within a hundred miles of our store. Better yet, it was organic and corn- and soy-free. That was important to us, because these crops do not grow well in this area and must be transported over long distances to us, at great expense and use of natural resources. If you live in an area where these crops are grown locally, this aspect might not be so significant to you. We think it is most important to seek out the best feeds available and support your local farmers and businesspeople. Besides, the better the quality of the feed you provide, the better the eggs will taste!

Pasture and Insects

As we noted earlier, chickens need more than manufactured feed to thrive. Hens (and their eggs) are at their best when they have some access to your yard in order to graze and hunt. Chickens are omnivores and will happily eat a wide range of foods, including meat (think bugs) if given the opportunity. As we mentioned in chapter 1, they are closely related to dinosaurs, and we believe this is the basis for the fierce insect-hunting behavior we've witnessed. Ideally, your chickens would spend the entire day ranging around your (safely fenced) yard grazing your (chemical-free) lawn, unearthing worms and insects, and nibbling on weeds and other plants. They seem to have a pretty good sense of what's safe to eat, so they need no supervision besides whatever you are doing to keep them away from your prized hostas. If for any reason this is not practical, try to get them outside to graze and hunt for a couple of hours a day or on weekends. Whatever you do, be sure they return to the safety of the coop each night!

If there is simply no way to get the hens out of their run often enough, you should bring the greens to them. Whether it's weeds from the garden or wilted salad greens from the grocery store, your birds will appreciate it immensely and their egg quality will improve noticeably. One caution: very short, fresh grass clippings are fine, but anything longer than about ¾ inch can potentially cause serious digestive problems.

Grit and Shell

In addition to eating bugs and chomping grass, your hens will do something unexpected while outside: they will eat rocks! No, they don't crunch on boulders, but they will pick up tiny stones (called grit) and swallow them whole. They do this because they don't have teeth (hence the expression "scarce as hen's teeth"). The stones find their way into a muscular organ known as the gizzard, where they help to grind food into a finer consistency that aids digestion. Your birds will usually find plenty of appropriate stones to swallow, but if you have a stone-free yard or keep the chickens confined full-time, you will need to supplement the grit as part of their feeding routine. Grit is available in two sizes from most feed stores: #1 for chicks and #2 for adults. Simply scatter a pound or two once a month and let them find it. Alternatively, you can leave a small amount in a bowl for them to eat freely, as described shortly.

Laying hens use large amounts of calcium to form their eggshells. If they are on layer feed, they will generally be getting enough of this mineral. However, some individual chickens, especially heavy layers during warm weather, need more. You'll know this is the case if their eggs have successively more delicate shells. Although some sources recommend feeding hens' own eggshells back to them, we avoid this practice, due to our concern about hens eating their own eggs (see page 174). As you can imagine, this habit is difficult to reverse once the chickens discover how delicious their own eggs are. We much prefer a crushed oyster shell product, which has the additional benefit of adding marine minerals to your hens' diet.

Shells can be scattered on the ground, mixed with feed, or freely offered in a bowl. Scattering on the ground is fine if the feeding area is relatively clean and dry. Mixing directly with their feed usually works well unless your birds are finicky (mix about 1 part shell to 25 parts feed). Free feeding, in which the shell is offered at all times and consumed separately from other food as desired, is probably the best approach. The only drawback is that the hens will inevitably knock over any bowl you try to put the shells in, defeating the purpose. Try this instead: nail an empty tuna can to a heavy board (use a long, rustproof, toxin-free nail), elevate it about six inches off the ground to keep it clean, and refill as needed. You may want to add a second can for your grit while you're at it.

Scratch and Treats

The term "scratch" usually refers to any type of cracked or whole grains fed to poultry. Scratch should be considered a treat and offered in much smaller amounts than feed, because the high carbohydrate content of scratch can be fattening. Several formulations may be available; the most common is straight cracked corn. This also happens to be the sort that's highest in carbohydrates and lowest in protein. Because of this trait, we like to use corn in the winter—the easily digested, high-sugar grain helps to keep our hens warm at night and well insulated with a light layer of extra fat. In the summer, however, we suggest that you switch to a slower-digesting mix of grains like wheat, barley, or oats to prevent potential overheating. Better yet, simply eliminate scratch altogether during this season of abundant forage. Many of our customers have reported scratch "addiction," so consider phasing the scratch out slowly lest your birds need rehab.

The other kind of treat comes from your kitchen. You can toss your birds any type of grain-based foods, dairy products, or most vegetables and fruits provided they are not rotten (stale is fine). You can probably safely feed small amounts of well-cooked meat, though we avoid feeding chicken to our chickens because that's just creepy. There is some controversy about feeding chickens terrestrial meat for safety reasons, so it may be best to avoid all flesh except fish, which is quite healthful. More items to avoid include avocado, chocolate, citrus, onions, garlic, potato peels, salty foods, artificially sweetened foods, processed foods, spicy foods, and fatty meats.

Keep in mind that kitchen scraps and treats in general should be fed sparingly, as they will dilute the benefits of your feed and may lead to obesity. Although we admit that there's a certain appeal to the image of a yard full of plump hens, being overweight is really not good for their health. In addition to heart disease, excessive fat puts them at risk for a potentially fatal complication called egg binding, in which the hen cannot lay her egg because the oviduct is stopped up by layers of fat. You can read more about this sad fate in chapter 7 (page 148).

A final cautionary tale about treats: We have had several confused customers come into the shop telling us that their hens are simply not interested in their feed anymore. I have learned to ask them what else they are feeding them. Every time the answer is the same: human food and lots of it. One recent customer was sharing whole loaves of her almond bread; another was actually cooking meals for her birds. Of course the birds were uninterested in their chicken feed! Too much human food is just that—too much. It's simply not nutritionally complete enough and will distract your birds from what they should be eating.

Feedings

Now that you know *what* to feed your flock, let's explore *how* to feed them. As you will see, it's not enough to just leave out a full feeder and hope for the best.

Quantity

Knowing how much to feed your home flock is more of an art than a science. It changes based on the season, availability of forage, your flock's health, and your birds' ages, among other variables that are hard to account for. Don't worry, you will eventually get a feel for it. We have found a rule of thumb that works for most situations: give each bird a double handful of feed (about ½ to ¾ cup) and no more than a single handful (about ¼ to ⅓ cup) of scratch or treats a day. This is based on an average-size adult hand, so adjust as needed.

Alternatively, you can freely offer the birds as much as they can eat, but again, this may lead to obesity problems and can attract rodents and other pests who will gladly steal unattended food. For this reason, we suggest purchasing and using a large adult feeder, but filling it with only as much as will be consumed in the morning. If you will be going away for a couple of days and the coop and run are secure enough, you can fill the feeder to provide food over a longer period.

We increase the amount of feed we give our chickens in the winter because they need extra energy in the cold and there is less green stuff around for them to eat. Conversely, you will want to cut back a little in the warmer months. When in doubt, pick up your hens and heft them: they should feel dense but not heavy. If they feel too light, or seem to be laying lightly, give them a bit more feed, but don't overdo it.

Rats and Mice

Nearly every home and neighborhood in the country harbors at least a few rats or mice. The food you set out for your chickens also happens to be an ideal ration for these small, opportunistic mammals, potentially leading to explosive population growth. We are happy to say we have found a solution, based on our own experiences.

Soon after we started keeping chickens, we began to occasionally see a rat near the coop. It would emerge briefly from its hiding place, making regular visits to the chicken feeder, and disappear again. It did not bother the chickens nor their eggs, but nevertheless we did not want to play host to a large family of rodents. At first we tried a humane trap, but the rat seemed uninterested. Finally, we sought to address the problem at its source: the feed.

We decided to put out only as much food as our flock could consume in about two hours (see feeding recommendations). In our case, this meant giving them only about 3 cups each morning, rather than leaving their feeder full at all times. We adjusted the amount upward in cold weather and sometimes provided a smaller afternoon snack. If we left on an occasional short vacation, we filled the feeder more. The result? We have not seen that rat or any others since!

If you want to try other control methods, we suggest using carefully positioned traps (placed where the chickens can't be harmed) rather than poisons. Though effective, poisons are potentially dangerous to your chickens and other animals that may accidentally eat the poisoned grains (or the dead rat!).

Using the Feeder

It's advisable to hang the feeder from a chain or otherwise keep it off the ground to prevent your hens from making a mess of it or scratching the feed out onto the ground and causing further waste. Also, make sure it is in an area that is protected from rain but not inside the coop (as we discussed in chapter 5). Wet feed will get moldy, and if this happens, it should be discarded immediately unless it's consumed the same day as a wet mash (sometimes offered to finicky or stressed birds).

Galvanized hanging feeder (left) and fount (right)

Feed Storage

Feed is relatively expensive, so it's worthwhile to invest in a good storage container to keep near the flock or inside the house. Whatever you choose, it's very important to keep the feed dry and protected from scavengers and rodents. We use a small metal trash can with a securely fitting lid that we keep next to our coop. Secured in this simple bin, our feed stays dry and fresh for up to three months, despite the rain that we are so famous for here in Portland. It may be better to use a plastic container in areas of high humidity, because metal cans can "sweat" as humidity condenses inside them. Locate your feed container in a shady area, because heat, as well as moisture, will degrade your feed rapidly.

Feed storage container

You feed storage container must be secure from all manners of animals that would love to dine on your feed. Chief among these are rats, mice, and raccoons. Although we do have raccoons in the area (once we actually woke up to one peering in our bedroom window) and have seen an occasional mouse, they have not been able to get into our container. It's a good idea to use bungee cords or other methods to secure your lid in case it's tipped over by a hungry bandit.

Watering

Clean, drinkable water is even more essential for your chickens' health than food. Chickens do drink an impressive quantity of water, and we recommend investing in a fount that can supply a few days' water. You'll still want to check it often and add more if needed, swirling it around and dumping accumulated debris at the same time. Every week or so, dump it all out and change the water completely. Periodically, scrub it out well to remove any algae or other contaminants.

Because elevating your fount will help keep it clean, many customers ask about getting a hanging version. We have not yet found one we like that will resist freezing and thawing, UV rays, and rough handling. For this reason, we

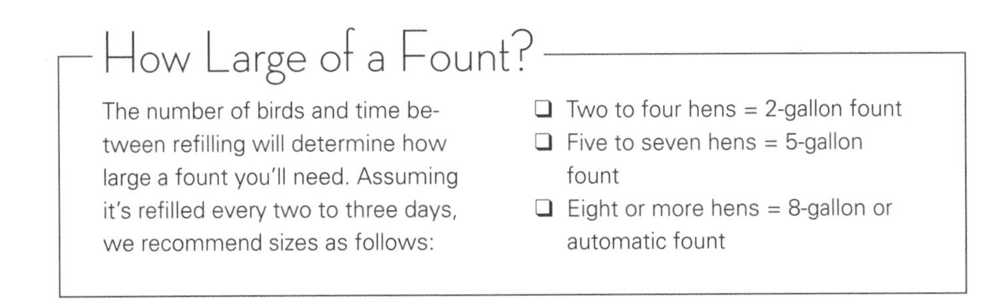

How Large of a Fount?

The number of birds and time between refilling will determine how large a fount you'll need. Assuming it's refilled every two to three days, we recommend sizes as follows:

- ❑ Two to four hens = 2-gallon fount
- ❑ Five to seven hens = 5-gallon fount
- ❑ Eight or more hens = 8-gallon or automatic fount

like the double-walled metal ones that, unfortunately, can't be hung. Instead, place this type of fount on two layers of two bricks placed side by side (that's four bricks) or some other stable, flat-topped object that will raise it up 6 to 8 inches.

It's much easier to determine how much water to provide than how much feed. The simple answer is "Give them all they want." Of course, you'll need to make sure that it's fresh and clean, but otherwise the only thing to know is that chickens can drink quite a lot. In summer, a single hen can easily drink a quart of water per day by herself and sometimes more. It's also important to remember that your birds will need fresh water even on rainy days. They will seek out and use puddles to drink from, but these smaller, more stagnant water sources may not be sufficient for their needs.

SANITATION

We covered this important topic briefly at the beginning of this chapter with a promise of more detail to come. The key ingredient for keeping your coop in a low-odor, sanitary condition is litter (aka bedding). The purpose of litter in the coop is twofold: (1) to provide a layer of material over the floor of the coop that can be removed periodically along with your chickens' droppings and (2) to help keep the interior of the coop, and the droppings inside, relatively dry by absorbing moisture. This is how litter helps to control "fowl" odors.

As with chicks, our favorite choice for the floor of the coop is pine shavings. They are very absorbent, control odor well, and work perfectly for the deep litter method mentioned on page 99. We do not recommend straw or hay for small urban coops. This is because most of these coops tend to be poorly ventilated; this causes these types of bedding materials to mildew, which can lead to upper respiratory problems in your flock. Worse yet, the hollow stems of the straw often harbor mites and other insects that can create a serious infestation that can be difficult to treat (see chapter 7). Unless you have a large coop in the country, we recommend that you stick with wood shavings (especially pine or aspen) for your coop's litter.

One place we will use straw is in the outdoor chicken run. Straw is ideally suited as an outdoor litter in muddy areas and is less expensive than wood shavings. We do not, however, suggest using straw in the nest box no matter how farm-appropriate it may seem. Instead, hay (the greener, more expensive bales) may be used in the nest boxes because it's less likely to harbor mites. Or simply follow our lead and use pine shavings in there as well.

Other options are available regionally. On the West Coast, many folks enjoy using rice hulls in both the run and the coop. Available in huge bags, these are quite attractive and inexpensive. The drawback to rice hulls is that they're lightweight, so wind and the chickens can easily spread it around, making quite a mess. Some of our customers also report rice sprouting around their coop after using hulls—we reassure them that the chickens will eat it before it causes a problem.

The Dirt on Litter

There are a variety of opinions about how often the litter in a coop must be changed. As we mentioned previously, we like to use the deep litter method, because it is very low-maintenance (are you sensing a theme?). Using this approach, we change our litter entirely only about four to eight times per year. We start with a generous layer of pine shavings, about 1 inch thick. As this gets soiled (and before it becomes too stinky), we add additional layers to it periodically. How often and how deep will depend on the size of the coop, the number of chickens housed there, and your climate, but a general guideline is to add about half an inch every week. When the layer is about 6 to 8 inches thick, or if odor is becoming an issue, it is time to scoop it all out. We recommend using a small rake for this job if you can reach into the coop with it. It's helpful to scoop the litter out onto a small tarp for transport to your compost pile or garbage. We have devoted a wet/dry vacuum to this cleaning task—it works great but is too soiled to use for anything else now.

In the alternative fresh litter method, you simply change out smaller amounts of litter more frequently. Some folks definitely prefer this; either they have not had luck with deep litter or they simply think the fresh method is better for sanitation reasons. The choice is yours, but you will likely go through a lot more litter this way and have a lot more to dispose of after it is used. If you use this method, count on cleaning every week to ten days.

Whichever method you choose, plan on doing a more thorough cleaning about every six months (more often if you have large numbers of birds). A sunny day is the best time to do this so the coop will dry more quickly after washing and spraying. For this cleaning, we start by scraping any hard surfaces that may have become crusted with poop or other materials. Then we remove all of the litter with a rake onto a tarp (into the compost pile it goes!). Next, we use a shop vac to remove all debris from the hard-to-reach corners of the coop. An optional step is to power-wash the interior of the coop if there is a lot of crusted material, or you think you might have a mite problem. Then we spray all surfaces with a mild bleach solution and, once it is dry, we sprinkle poultry dust (or diatomaceous earth dust) liberally, especially in corners and cracks. Finally, we lay down a new 1-inch layer of litter.

Composting Litter

If you use the deep litter method, you will likely discover that the bottom layers of soiled litter have already begun composting by the time you remove it. The whole load of litter can be added directly to your compost bin and is a great source of carbon (from the litter) and nitrogen (from the poop). If our compost pile is too full to accommodate the whole load, we spread the rest of the litter under our fruit trees as a mulch. Although it is true that fresh chicken poop is too "hot" to be applied directly to vegetables, the small amounts mixed up in the litter will be quite harmless. Pure poop from a droppings tray, which is a more concentrated source of nitrogen, definitely does need to be composted or aged before it is applied to any of your plants.

Straight pine shavings and those from the fresh litter method will be a little harder to compost. These are best mixed with other material high in nitrogen such as food scraps, vegetable peelings, manure, or green grass clippings. The carbon in the shavings will help balance the pile, and the heap should be ready for the garden in a couple of months. Have you had a problem with overly "juicy" or smelly compost? Chances are you were in need of extra carbon anyway.

PROTECTION FROM PREDATORS

The final ingredient for your chicken care system is protection. As we discussed in the coop chapter, your girls will need to be protected from predators, weather extremes, and their own desire to investigate what they are sure are greener pastures on the other side of the fence. Having a fenced yard and a well-built coop

Better than TV

and run will go a long way toward fulfilling all these needs. Let us now look at a few more things that you can do to further tip the odds of survival in your direction.

Your coop, and to a lesser extent your run, should be a predator-proof sanctuary for your flock. If you have built your coop from the plans in this book, you are probably in fairly good shape. We would like to reiterate that the coop must be capable of being fully secured at night. But for that to work, you also must remember to shut the hens in after dark. We suggest setting a regular alarm on one or more watches, or programming a reminder text message sent from an online calendar. After a few months you will not need the prodding, but it's very helpful to set the habit. If you think it's going to be a big problem to remember to lock them up every night, or if you're not home around dusk, there are electric chicken doors that operate on timers or light sensors. We have invested in one of these expensive little gems ourselves, and we appreciate the secure feeling that it gives us on the occasions when we are unable to return home at the usual time.

Cats and Dogs

Generally speaking, cats are not a major threat to chickens. A full-grown standard-size chicken is large enough to intimidate an average cat, especially when it has the strength of numbers in its flock. However, if you have chosen to raise bantam breeds, or your birds are not yet fully grown, they may be safer enclosed in a run unless you are around.

Dogs can pose more of a threat, depending on the dog and the circumstances. Years ago, Hannah experienced a situation in which neighborhood dogs who were roaming the streets on their own jumped over her fence and attacked her chickens repeatedly. It was an unforgettable tragedy enabled by a surrounding fence that was only 4 feet high—not nearly high enough to keep out these motivated dogs. In another instance, Robert was taking care of a family member's dog when she escaped into the backyard and killed one of his favorite hens. This is a good example of what we call "the inside job" and is one of the saddest ways to lose a chicken.

Happily, our own two dogs think chickens are the most boring things in the world. They happen to be Labradors, which are bred to fetch but are not usually big hunters of small prey (unlike many other breeds, such as terriers). In our case, after many supervised interactions, we decided the chickens and dogs were safe with each other, and all has been peaceful for over two years now. Please proceed with caution in making a similar decision. Some of our customers have told us that their dog was fine with the chickens until one of the chickens started to run for some reason. At that moment, the dog seemed to think this was a game, chased after the chicken, and shook it playfully—just hard enough to break its neck. The dog clearly did not intend to kill the chicken but was unable to resist the urge to chase. Bottom line: Dogs and chickens *can* coexist peacefully, but not in all cases.

Some dogs definitely do see chickens as prey, and their owners are wise to keep their chickens confined to their run all day or at least whenever their dog is near. They should also make sure that the run is sturdy enough to resist the impact of a dog running into it at full speed. Only welded wire or hardware cloth on a sturdy frame will do the trick.

Coming Home to Roost

Each evening around sundown, your birds will instinctively return to the coop, hop up on their roosts, and settle down to sleep. You merely need to close the doors behind them to keep them snug and protected. That's how it works in theory, anyway. The reality is that some birds do instinctively do this, and others need some help from their parents (that's you!) to get them into the routine.

If you have recently moved your birds outside, added new birds to the flock, experienced a sudden seasonal change, or added lights to the coop or yard, your birds may instead perch outside the coop (or even farther away) without entering it as expected. These birds are now exposed to the elements and predators and must be trained or retrained to find the safety of the coop. Fortunately, it's not too hard.

Buddies flock together

At first, collect the birds one at a time (they should be sluggish and easy to catch) and place them inside the coop on their roost and shut the door. Do this for two or three days if needed. Then, for the next couple of days, place them inside the entrance of the coop and close the door behind them. If they still don't get it, try putting them at the entrance and walking away, returning to close the door a few minutes later. By gradually placing them farther from the roosts, you will be teaching them to move toward the coop as the daylight wanes.

Flights of Fancy

Beyond protecting chickens from predators, you will also need to protect chickens from themselves. Although they reliably return to their coop at night (assuming they can find it), throughout the day chickens will roam far and wide if given the opportunity. In an urban setting, this makes them vulnerable to getting lost, run over by cars, or attacked by neighborhood dogs. In general, the same precautions you are already taking to protect your chickens from predators will also protect them from escape. Keep in mind that different varieties of chickens have

Cutting flight feathers

varying abilities to fly, and you might want to keep the height of your fence in mind when you choose your breeds. Most standard or "heavy" breeds cannot catch much air (generally 4 to 6 feet), but we have heard that bantams can easily fly up to perch on tree branches.

If your chickens are making you nervous with their demonstrations of aerodynamics, consider clipping their flight feathers. It is usually recommended to clip only one side, because this limits the chicken's ability to fly in a straight line, but Robert likes to clip both wings to reduce overall lift. Hannah was especially reluctant to mar the beauty of her birds this way. Although she was aware that there are no nerves in feathers so this would cause no pain, she was worried that it would make the birds look like they'd had a crew cut. Once she bit the bullet and did it for one of her more tenacious escape artists, she was relieved to discover that the change was virtually invisible because these feathers were usually tucked under the wing.

When clipping wing feathers, you actually need to trim only the three or four largest feathers to impair flight. The best time to do this (or any other intervention or examination) is at night, when your chickens will be docile, easy to catch, and essentially unaware of what you are doing. In general, nighttime is a great time to perform any procedures on your chickens that you think would usually make them struggle. We'll talk much more about this sort of activity in our chapter on chicken health.

DEALING WITH THE ELEMENTS

Inclement weather, not just other animals, can be harmful to your flock's well-being. Fortunately, most chicken breeds, with their layers of insulating feathers that help moderate both heat and cold, are built to withstand all but the nastiest weather. In most climates, there is really not much a chicken keeper has to do to keep their birds comfortable. However, in areas of weather extremes, you should be prepared to help them out.

Cold

If you are expecting temperatures just a little below freezing, you probably won't need to take any special precautions for your flock. The exception would be bantams (half-sized versions of popular breeds) or other small breeds. Because of their smaller body size, these birds should be provided with supplemental heat at temperatures or wind chills of 32°F or below. Other breeds should be hardy to about 20°F, plus or minus depending on wind, coop quality, and number of birds in a coop. If called for, supplemental heat can be offered in a number of ways, which we describe shortly. It is a good idea to close all doors to your coop overnight at all times, but especially when temperatures dip below freezing.

If you live in the upper Midwest or another exceptionally chilly climate, you probably want to use a variety of approaches to handle the cold: choose exceptionally cold-hardy breeds (such as the Rhode Island Red, Plymouth Rock, Brahma, or most any other non-Mediterranean breed), build a very sturdy coop, and provide some supplemental heat during cold parts of the year. Although the bodies of hardy breeds are nicely insulated by their feathers, the combs and toes are susceptible to frostbite in extreme cold, so you will need to keep the birds warm.

There are a number of ways to provide supplemental heat. The most common approach is to reuse your brooder reflector and bulb that you got when the chicks first arrived (assuming you started with chicks). If you use this or another appropriate lamp, we recommend using a red bulb that will allow the chickens to sleep better while it's on. When mounting the reflector, be sure to use screws or otherwise secure the lamp so it can't fall into the litter. Also, try to mount it about 3 feet above where the chickens will be roosting on the roost bar (where they will be sleeping). Any closer and the birds may be singed or overheated.

Another approach is to use an outdoor-rated pet heat mat normally used for keeping Fido from freezing his tail off. It releases a relatively small amount of heat, but you can place it directly below the roosts (on the floor or poop board if you're using one) where they need it most. You can try other heating devices as long as they are not a fire hazard in any way and do not create an irritating draft. This rules out most home heaters. If you do use an electrical appliance, it's a good idea to also use a timer or thermostat to ensure that the heat is on only when needed.

Cautions: With all electrical devices, be certain to use only outdoor-rated appliances and extension cords. Do not plug more than one appliance into a single cord; plug each appliance into its own outlet. When installing devices inside the coop, be careful to secure them in such a way that they will not fall into or otherwise contact the litter and pose a fire hazard. Do not combine blankets or other tight coverings with electrical heaters, as this could lead to overheating and even fire.

Low-tech approaches can include traditional hot water bottles, a hot compost pile under the coop, or any number of other passive strategies. Some people also add temporary insulation inside the coop in the form of straw (I know we said not to, but if that's all you have, it's better than letting your birds freeze) or old blankets. Throwing a tarp or blanket loosely over the coop may also serve to retain some heat and should help to block winds.

Providing water in the winter is trickier if you live in an area where the temperatures regularly drop below freezing. Perhaps the simplest approach to light freezes is to bring the fount indoors at night and return it thawed in the morning, replacing it with more thawed water in the afternoon. Don't worry about

Close Your Doors

We are often surprised by folks who don't close the door to the coop at night. This simple step will save your birds from both freezing winds and predators, and doing it manually yourself costs nothing. If you have no door to the coop, consider propping a board against the opening when the weather becomes nasty.

your chickens getting thirsty at night—they are so sedate and immobilized in the dark, they won't even notice that their water is missing. Another approach in very cold climates is to keep the water slightly warmed by using special heaters available for this purpose. If at all possible, try to avoid bringing the fount into the coop to warm it unless the hens must be temporarily confined there, too. As we mentioned in chapter 5, cross-contamination resulting from poop getting into drinking water is the source of many health problems for chickens.

All of these interventions are most critically important at night when the birds are sleeping and not moving much. During the day when they become active, your hens will withstand cold better, and they will be able to return to the heated coop or otherwise seek shelter if needed. As mentioned earlier, a sleeping hen is most susceptible to frostbite on her comb and feet. Because her feet can be tucked under her body, they are usually protected from freezing air and need little care beyond providing a broad roosting bar. Her comb, on the other hand, is exposed to the air and benefits from a coating of petroleum jelly in extreme conditions to help conserve moisture and heat and prevent cracking. (We would recommend nonpetroleum alternatives like vegetable oil or lard, but that delicious coating may encourage other hens to take a tasty peck!)

As mentioned in the feed section, a simple way of protecting your birds from chilling overnight is to send them to bed with a full crop. If you give them an extra handful each of scratch grains like cracked corn in the afternoon, your hens will get a boost of high-carbohydrate energy to burn through the night. If practiced regularly, this is a good way for them to put on a moderate layer of fat during the winter that will act as insulation.

Heat

Very hot climates (think Texas) present the opposite problem, and you need to make sure your birds are protected from heat stroke. When it is hot, you will see your chickens keep to the shade and pant with their beaks open, walking around with their wings held away from their bodies. This is how they cool themselves, much like your cat sprawling on a cool tile floor, or your dog panting in the shade, and is not cause for great concern.

You will also notice that they will drink more water on scorching days, so your primary intervention will be to make sure the fount is kept full of cool water and moved to a shady spot. If it gets really hot, you may wish to offer a muddy area to wallow in or otherwise wet down your chickens. In the most extreme cases you may even consider moving your birds to the cool of your basement or into an air-conditioned room. In general, providing plenty of fresh water and some shade will get these formerly tropical birds through all but the worst heat waves. If you know that heat will be a yearly and severe threat, consider choosing less heavily feathered or more heat-tolerant breeds (such as the Brahma or any of the Mediterranean breeds) from the outset. Breeds from the Mediterranean and other hot regions tend to do well in similar climates around the world.

INTRODUCING NEW BIRDS

There comes a time in almost every chicken keeper's life where they must introduce new birds to their established flock. Perhaps you have lost a few to predators and are replenishing their numbers. Or you may have been able to purchase a breed that you have long wished to own and want to introduce them to your flock. Whatever your motivation, adding new girls to your flock can be a difficult process because of their social structure. The "pecking order" is a very real thing; in this case, it means that established birds in a flock will try to dominate their new mates. This can lead to aggressive behavior and even injury.

The first thing to know is that a single chicken being introduced to an established flock is bound to have a hard time of it. This lone addition will be treated as a threat by the other, established hens; they will repeatedly bully her. Sad to say, some "new girls" may never be fully integrated into the group, though most will eventually. Introducing two or more adult or older juvenile birds to a flock tends to be a much smoother process. These new birds will upset the flock enough to force the resetting of the pecking order, leading to a new equilibrium that includes the new birds. Avoid adding birds under the age of eight weeks or so to an adult flock because they are generally not able to defend themselves. Ideally, young birds should be ten to twelve weeks old before you introduce them to an older flock.

Several sources suggest introducing new birds by providing them with a separate cage or sectioning off part of the coop for them with their own food and water. If you allow old and new birds to interact safely through the fencing, the theory goes, the birds will become accustomed to each other gradually, reducing conflict. Although this works for many, we feel that this approach is time-consuming and potentially expensive, and merely delays the inevitable day that the birds must come together and sort out their places in the pecking order.

This approach does have the advantage of allowing the newly introduced birds, if they are young, to spend a couple of weeks outdoors while they continue to mature. It also allows the integrating pullets to complete the recommended period of eating starter feed (potentially medicated and not recommended for feeding to laying hens) while the rest of your flock consumes their regular layer ration.

A compromise that has worked for us at the store is to provide a hiding and sheltering area in the run or coop, such as a cardboard box with a 6-inch doorway that's too small for the adults to enter. This is also a good location to place medicated starter feed to keep it away from the adults.

Alternatively, you can offer starter feed to the whole flock along with oyster shell to provide calcium for the layers. We suggest this approach because the extra protein in the starter will not harm the adults as the calcium in the layer feed would the chicks. The only downside to this strategy, as noted already, is that most starter is medicated and could temporarily make the adults' eggs unhealthful to eat. Our solution? Forgo eating the eggs for about two weeks past the last feeding of medicated starter (when the new birds have reached twelve weeks of age).

However you approach the feed and shelter situation, we recommend integrating the new birds into the flock by placing them inside the coop, on the roost after nightfall so that the initial introduction occurs during their sleepy, slow period of activity. Then, first thing in the morning, let them all out of the confined space into a large, open area—at a minimum into the run, and ideally into the yard. Here's the best part of the strategy: provide them with an activity that will distract all of them to some extent. We like to employ the seduction of widely scattered scratch grains—they spend hours looking down, searching for these tidbits, and seem to hardly notice who has joined them.

Lastly, it bears repeating that younger birds are still vulnerable to predators, such as cats, that adult birds can easily handle. For this reason it's worth considering confining the new little darlings to the coop and run area while the adults free range.

CHICKENS IN THE GARDEN

The concept of protection also extends to your chickens' relationship with your yard and garden. We are often asked whether chickens can be harmed by eating poisonous garden plants. Although they are not the smartest animals, we have noticed that chickens have some instinct to avoid bitter and strongly aromatic plants that could be dangerous to them. Nonetheless, it would be wise to avoid planting anything known to be hazardous in the area frequented by your hens. Likewise, you should not store garden or household chemicals, even organic ones, where the birds can get into them. Use of these substances in accordance with their labeled directions should be safe in most cases, but when in doubt avoid using them where the chickens are kept.

As we mentioned in chapter 1, chickens can potentially be quite destructive when it comes to tender garden plants and vegetables and will poop everywhere. Luckily, it's relatively simple to make a few changes that will minimize

Large run and fenced garden

their impact. To chickenproof our garden, we first designated our front yard as a chicken-free zone and focused our delicate annual and perennial plantings in this area. Segregated in that way, the birds simply had no access to the most easily damaged plants that grew at just the right level for them to eat. This left our backyard to be shared by both the birds and the rest of our garden. Here we plant all of our fruit trees and larger shrubs that grow well beyond the reach of our hens (they browse up to about 2½ feet). They have free access to the lawn and the mulched areas under established plants.

If you have plants that are attractive to the flock and must coexist with them, you will need to create physical barriers to protect them. The simplest and least expensive is bird netting. Practically invisible, black netting can be draped over strawberries, blueberries, vegetables and anything else that you think might be targeted. For us, netting was not practical to cover our extensive vegetable patch. We opted for fencing and soon learned that a 6-foot height worked best. To our delight, a wire field fence (not cyclone) with wooden posts and a gate actually enhanced the tidy appearance of the backyard and made a highly effective barrier.

We have found it helpful to arrange a few large stones or bricks around the base of newly planted trees and shrubs to protect them. Hens love to scratch at newly disturbed soil looking for tasty tidbits. Although fun to watch, this behavior will eventually damage the roots of new plantings, so some effort must be made to minimize it. This is seldom a serious problem for established plants.

Your patio or deck are other areas that you may want to fence off. In our case this was impractical, so we've learned to manage the consequences. By this, we of course are referring again to poop. We wish we had a keen idea to keep these outdoor gathering spots clean with no effort, but alas, we don't. Our solution: Robert uses a pressure washer to quickly spray off the patio. He says that it takes only a few minutes and can be fun, but the task must be repeated whenever company comes over or Hannah requests it.

You can channel the destructive nature of an active flock to do work for you. If you confine the birds in a fenced area, small pen, or mobile coop, their grazing and digging can be focused in one place until the land is cleared. We use this strategy in the fall to help clean up the remains of the vegetables. They love eating the mushy tomatoes, wilted greens, and fallen sunflower seeds. Better yet, there are many delicious and nutritious bugs around this time of year as well. In the spring we admit them into the vegetable patch again to graze on the cover-crop plants that we've sown over the remains of the previous year's garden.

Fertilizer

Another unexpected benefit of chicken-garden integration is that simple rule of farming: what goes in as feed comes out as poop. When it comes to chickens,

Myth: All Chicken Poop Burns Veggies

It's an accepted gardening truism that chicken poop will "burn" your plants. This is true to the extent that large quantities of pure (undiluted by litter), uncomposted chicken droppings can harm plants if applied thickly because of its high available-nitrogen content. This is not true of the stuff that you will collect and compost from a home flock, nor what they will distribute themselves around the yard. Because it occurs in lower concentrations, your material will be easily managed by the naturally occurring bacterial population in your yard and will be rendered harmless, and in fact quite beneficial to your plants, in the process.

this poop makes great fertilizer. In fact, of all manures, chicken waste is the most effective as a fertilizer if properly handled. This means that the 20 pounds (or more) of feed that you give your flock monthly is like a regular deposit of fertilizer into your garden's bank account. Better yet, your hens will gladly do most of the spreading for you. The rest can be composted along with other debris and then placed exactly where you want to use it.

PROPER CHICKEN HANDLING

At some point you will need to hold your chickens, and it is important to do so properly. Whether your intent is to cuddle, inspect, or wrangle the hen, if you hold her in a way that feels insecure, she will struggle and could harm herself in the process. Keep in mind that a tame hen is far easier to catch. All the socializing you did while your bird was a chick is about to pay off.

First, reach down and swiftly snatch your chicken around the middle, using both hands. This will allow you to pin down her wings, which prevents wild flapping. If her wings are not secure, the hen will make quite a scene, terrifying herself and slapping you repeatedly in the face.

Once you have your bird in your hands, we recommend shifting her to under the elbow of one arm (like a football), keeping her wings secure the whole time. In this position, she will feel "swaddled" and will soon become calm. You can support her belly, or even hold her legs from beneath, which will add to her sense of security. Once grasped in this way, you can gently manipulate her as you inspect for wounds, parasites, offer medical treatment, or simply stroke her on the breast and neck (favorite massage spots). When we need to move a bird, we simply tuck her under one arm, stroking her with the other hand, and walk off. A word to the wise: Make sure that the vent is kept behind your torso to prevent the dreaded "pocket poop!" (Also keep in mind that their claws are sharp.)

It should be noted that while some birds will crouch when you approach (in a sexually receptive manner) making them easy to gather, others will actively evade you. You have three choices: chase them into a corner, lure them with treats, or wait until after they've gone to sleep and pluck them from the roost at night. We usually rely on the latter method when needing to handle flighty birds like our Sicilian Buttercup.

Whew! That's a lot of information to process about adult chickens! Don't worry if you don't feel like you've mastered it all right now. Much of this will come to you instinctually as you tap into your long-repressed inner farmer. Refer back to this chapter frequently as you need help. Before long you'll develop your own routines and practices that fit your lifestyle and unique flock.

Crouching Chicken

You may notice that once your chickens reach maturity and start laying, they will suddenly crouch down whenever you approach. No, they haven't become frightened of you, they are simply demonstrating an instinct that is designed to allow a rooster to fertilize them. Once this instinct kicks in, your chickens will be far easier to catch. However, if your chickens are broody, molting, or otherwise not laying, you may notice that this instinct disappears temporarily.

7

HEALTH MANAGEMENT: SHOULD I BRING A $5 CHICKEN TO THE VET?

ONE OF OUR FIRST MAJOR DILEMMAS as chicken owners came shortly after one of our favorite chickens survived an encounter with a dog. Rosy was left with a truly shocking swath of exposed flesh; the missing skin included her entire tail. She was a pitiful mess, and we were pretty sure that she'd die from the wound. Robert wondered aloud if we should take her to the vet. It was a reasonable question—if it had been one of the dogs there would be no doubt that we would. But a chicken? We weren't sure about that. In the end, we decided to treat her ourselves, and, happily, she survived. In fact, she grew back all that skin, and somehow that skin knew it was supposed to have tail feathers attached. Looking at her rich plumage today, you would never know what she has been through.

At the other end of the spectrum, one of our favorite customers once spent a shocking amount of money to have an avian veterinarian give her chicken a life-saving hysterectomy. She certainly would never lay another egg, so this was clearly a chicken that had risen to full-fledged pet status. Luckily, with a little guidance, you can treat or prevent most common chicken health complaints like a pro—at a fraction of the cost of a vet visit.

PREVENTION

There are many things that you can do to protect members of your flock from becoming sick or injured. If you start by reading our Chick and Adult Chicken Care chapters (chapters 4 and 6), you'll have a good sense of the kind of things you must do to keep your birds healthy. It all really comes down to these three things: diet, protection, and sanitation. Let's briefly review these again, this time in the specific context of health maintenance. You might find this diagram helpful to decipher our jargon as we discuss various chicken parts.

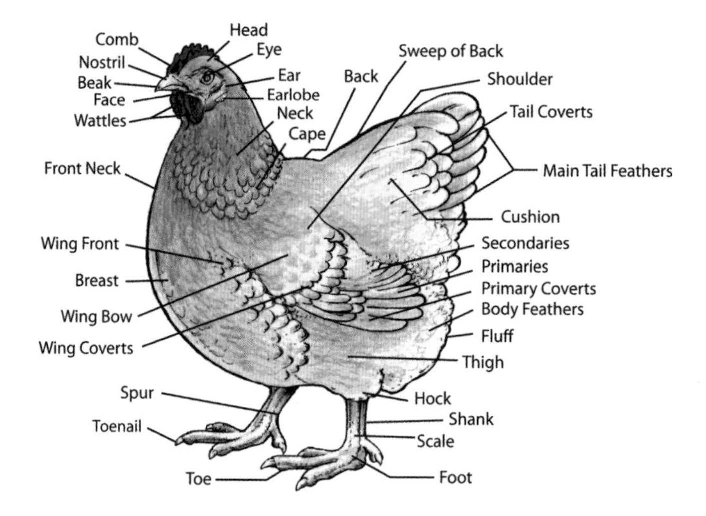

Diet

A wholesome, fresh diet that contains the right nutrients in the proper proportions is the foundation of good health for chickens (and people, too!). Feeding poultry properly encourages a vigorous immune system that is better able to prevent (and recover from) certain types of disease. To help ensure that chickens are getting what they need, we have found that it is worth the extra investment to purchase the highest-quality feeds available. Beyond that, they benefit tremendously from eating greens, whether from grass or leafy vegetables. Chickens also seem to have a special love for dairy products, so we suggest using them as a treat and protein supplement during times of stress, such as when they are molting.

Recent findings suggest that poultry benefit from the addition of probiotics—live microorganisms that live within a host in a mutually supportive arrangement. Some feed formulas now include these microbes; we suggest opting for them when possible. Otherwise, you can supplement with occasional treats of plain yogurt to obtain similar benefits.

Protection

As important as proper nutrition is to the health of your flock, it is all for naught if your birds are attacked by predators or injured by severe weather. Keeping the doors to a secure coop closed at night is probably the most important thing you can do to protect your birds from both of these threats. As suggested earlier, you should set a reminder on your watch or phone or do whatever it takes to remember to visit your birds at sundown and shut all access doors that you may have opened. Be vigilant against inside jobs as well: domestic dogs, if allowed access, may suddenly turn on your chickens. If you want to be sure that your birds are secure in areas of high predator populations, be prepared by building your coop and run to be a large, sturdy sanctuary where the chickens can spend most of their lives.

Providing shelter from the sun, wind, rain, or snow is also a must to keep your birds in good condition throughout the year. Covering some or all of your run with roofing material is one approach, and tarps can also be effective for temporary shelters. Build your coop to withstand the elements, but not so tightly that ventilation suffers. Poor ventilation coupled with poor sanitation (more on this follows) is a recipe for illness.

Sanitation

Although your coop need not be spotless (and few come close), there are a few important things you can do to keep things sanitary. First, don't locate the hens' water or food inside the coop. By placing these items in close proximity to the soiled litter found here, you are inviting cross-contamination. Besides, the coop is for sleeping and laying eggs only—your birds need not eat or drink at night or while on the nest. Instead, place the fount and feeder in a sheltered location inside the run. Place them on bricks or blocks, raising them 6 to 8 inches. This will help to prevent debris from being kicked into them—another major source of cross-contamination.

It's also important to keep your stockpile of chicken feed dry and protected by storing it in a trash can or other receptacle near the coop or indoors. Moisture and feed are a dangerous mix that can lead to the proliferation of disease-causing organisms and spoilage.

Lastly, keeping the coop and run reasonably clean will help prevent a number of health-threatening conditions (see chapter 6 for cleaning details). If you choose to use the fresh litter method, plan on cleaning the coop about two to four times per month. If you use the deep litter method, you will want to add a layer of litter weekly and muck it out every two to three months. Either way, plan on a more thorough cleaning about twice a year using a mild disinfectant and poultry dust (a powder usually containing mild insecticide synthesized from a naturally occurring substance found in the chrysanthemum plant). Rake the run at this time also.

Preventing the Spread of Disease

Diseases can be carried from one backyard flock to another on the soles of shoes or by sharing contaminated equipment. At modern commercial poultry farms, rigorous cleanliness is maintained by wearing protective clothing and dipping shoes in baths of sanitizer. Although we do not recommend that you go to these extremes when visiting a friend's flock, it is probably a good idea not to share equipment without scrubbing it clean and soaking it in a mild bleach solution. We've taken to leaving a pair of gardening shoes by the back door and wearing them when heading out to the coop. This way the shoes we wear to the store will have less of a chance of spreading any potential pathogens to our customers' birds and vice versa. If you are visiting a flock known to have a disease, consider wearing shoes that can be thrown in the washing machine on the hot cycle or dipped into a bleach solution.

It is even possible for some diseases to be spread from chicken to human. The most common of these is salmonella—bacteria that are commonly carried by poultry. For a human to contract it, he or she would have to ingest some chicken fecal matter. We don't suppose this is something the average backyard chicken keeper intends to do, but it is possible that this could happen inadvertently. For instance, chicks are adorable and appear to be highly kissable. Your children in particular might find it impossible to resist. Unfortunately, chicks do not discriminate about where they sit, and their fluff will invariably have traces of poop on it. Even handling your chickens and then eating a meal could be a means of transmission. We strongly recommend washing up after close contact with your chickens, and watching your children closely in their interactions with your flock. At the store, we keep a bottle of hand sanitizer next to the chick bins. You might consider this precaution for your home brooder. Salmonella will generally begin to cause symptoms one to three days after exposure. The most common symptoms include stomach ache, diarrhea, and fever, but people with weakened immune systems could get severely ill, even necessitating hospitalization. For this reason, the Centers for Disease Control advise that people who are immunocompromised and children under five years old not handle young birds.

The other illness of note that could be passed from bird to human has so far been the source of much worry without much actual consequence. We are speaking here, of course, about avian influenza. As of yet, the fears that it could mutate to be contagious from human to human have not become reality. The people who have contracted it are generally those who have contact with infected birds (usually in large numbers) in the course of their daily work. In addition, it has not yet been found in the United States. If you become concerned about the risk

Dust Bathing

To prevent parasites, your hens will naturally attempt to "dust bathe." Hens adore few things more than settling down in some dust (or almost any fine material) and vigorously pumping their wings in such a way that it flies up through their feathers. It is lots of fun to watch; even better, it makes it very difficult for parasites to survive. If your chickens have access to your yard, they will soon find a dry area in which they can dust bathe. If they are confined to a run, provide them with a basin full of fine dry soil or sand.

of avian flu, the best approach is to keep your flock away from wild birds, which could bring the illness from other regions. If your chickens spend all their time in a covered coop and run, they will be protected from any potential exposure, and as a result, so will you.

Don't Panic, Be Informed

We advise only skimming the rest of this chapter unless you have specific concerns about your flock's health. We say this because we have found that offering too much of this type of information up front can feel overwhelming for those just starting out with chicken keeping. We'd like to stress again that most of these issues are really quite rare and you are unlikely to encounter many of them during years of chicken keeping. On the other hand, we want you to have the information at hand if an issue does arise.

In addition to this book, you will want to have some local resources available should you need help quickly. First, find the name and number of a good vet in your area who will see poultry. The Association of Avian Veterinarians has a useful search option on its website, www.aav.org. Next, see whether there is a local chicken-keeping group in your area. In Portland there is a very active backyard chicken forum, and folks there are always throwing out questions, answers, and helpful hints. We bet that once you are connected with such a community you will find someone in your area who may not be a vet but is passionate and highly self-educated when it comes to chickens. We have just such a person in our area: Tonya, a beloved local chicken enthusiast who has been invaluable to us and our customers on many occasions.

FUN FACT

The world's oldest chicken, according to the Guinness Book of World Records, died of heart failure at the ripe old age of sixteen years old.

It's wise to give some thought to what your approach to chicken health care will be before a problem ever arises. Are you okay with incurring a substantial veterinary fee for care of your chickens? For most of you, this will depend on where your chickens fall on the pet-to-livestock spectrum. We find that most backyard chicken keepers consider their hens pets, but not quite at the same level as their dogs and cats. Others love their chickens without qualification and will do anything to come to their aid. It's also important to consider end-of-life issues before they occur: are you comfortable allowing a chicken who cannot be cured to die naturally, are you prepared to euthanize her yourself, or will you want a vet's assistance with this? Unless you are raising chickens for meat, the idea of slaughtering and eating your doomed chicken may well be repellent—but if the idea seems reasonable to you, just make sure the chicken in question is dying from an injury, rather than a disease that may, at best, affect the quality of the meat, or at worst, cause you harm.

Many local feed stores can help you with simple diagnosis of common ailments. They may also sell medications that will help, usually at a fraction of the cost of getting them from your vet. No matter how sure they seem about your situation, remember that they are unable to offer a proper veterinary diagnosis when it come to more serious or complicated issues.

GENERAL PRINCIPLES OF CHICKEN HEALTH CARE

There are many potential chicken health problems, but only a few that you are likely to encounter during a lifetime of chicken keeping. For this reason, the rest of this chapter will separate common from rare concerns in each category. We discuss the most common ailments first, as they are the more likely culprits when you are attempting chicken diagnosis.

The main thing we've learned from our fellow chicken keepers is that it's often difficult to conclusively diagnose a chicken, even for a vet, because of their unusual physiology and quirky behavior. As a result, we frequently get calls from customers whose hen has a limp or other vague symptom and they are worried that it may indicate something more serious. We usually suggest observing the bird for a couple of days to see whether the problem is getting worse or resolving itself before taking her to the vet. Often, taking the bird inside and feeding a mash of feed mixed with milk or water for a couple of days will allow the bird to get a leg up on the problem herself. This form of isolation may also protect the rest of the flock if the disease is contagious.

If things are not resolving, antibiotics are the recommended treatment for a number of poultry health conditions. Still, we recommend against using them willy-nilly every time your chicken sneezes. Many illnesses are caused by viruses, and antibiotics will have no effect in this situation unless there is a secondary

Hydration

The most common acute consequence of poultry disease is probably dehydration—it's important to do your best to get some fluids into your parched hen! A natural choice for forcing fluids is an eyedropper. If you try this, be very careful not to squirt the water down the wrong pipe. The safest method is to put a drop on the side of the beak and repeat as each drop is naturally swallowed. This technique is particularly effective with chicks. Our preferred approach is to feed a wet mash of water or milk mixed with feed, if the patient still has an appetite.

infection present. Also, as in human antibiotic use, overuse can lead to creating more resistant organisms that are tougher to treat. In addition, if your hens are of laying age, you will have to discard any eggs they lay for a period of time (usually specified on the medication's labeling). Finally, antibiotics may also target intestinal bacteria that aid with healthy digestion, leading to diarrhea and dehydration. (As mentioned earlier, adding yogurt to the chicken's diet during and after antibiotic administration may prevent a die-down of beneficial bacteria—and your chicken will love it.)

That said, there are situations in which antibiotic use is completely appropriate. If you think your chickens are in danger, and antibiotics are likely to be lifesaving, proceed by all means. Most antibiotics that are used for poultry are administered by dissolving them in the chickens' drinking water at a prescribed concentration. A chicken will generally drink a predictable quantity of water each day, and the concentration in the water is designed to deliver the dose she needs over the course of a day's water intake. A variety of antibiotics are available at feed stores or at online veterinary supply stores. Poultry electrolyte powder is also available in most feed stores, and it's safe to use this form of tonic liberally when a chicken is looking a little under the weather. It will make the water look like lemonade, but will help keep your patient well hydrated.

Vaccines are available for many of the most serious potential chicken ailments. The most commonly used vaccine is for Marek's disease; again, you should make sure any chicks you purchase are vaccinated. A vaccine for coccidiosis is also available now at some hatcheries if requested, and we recommend it heartily; although there is some controversy about the use of medicated feed with vaccinated chicks, this presents a problem only if you have a mixed batch of vaccinated and unvaccinated chicks. If you have experienced other tragic infectious outbreaks in your home flock, and you know which organism is at fault, it is often a good idea to vaccinate against that particular bug if a vaccine is available. Your vet will help you both with diagnosis of your past woes and with administering vaccines.

In most cases, a very sick hen will benefit from being removed from the rest of the flock. This will protect her from the picking and bullying that a chicken may be subjected to when she isn't feeling her best and thus appears weak. Even a chicken that previously occupied the top tier of the pecking order will suffer a precipitous (but likely temporary) plunge in her prestige level. Isolating the affected bird will also allow you to keep her in extra-sanitary conditions. An ill chicken usually will not perch anyway, so a confined area in protected surroundings, such as a garage or shed, will probably serve the purpose. If the weather is cold, you should provide extra heat for your patient. She needs her immune system to have as few external stressors as possible. If her illness is infectious, you want her away from the other birds anyway, to prevent the disease from spreading. If she is in need of medication, an isolated chicken can be dosed without having to treat the whole flock.

Chicken Health Assessment

Chickens will generally let you know when there is something wrong with them. For this reason, don't feel obliged to give your chickens full physical exams on a regular basis. If they are contentedly going about their business, laying eggs, and looking perky, leave them be. Keep your eyes open for substantial changes in behavior or appearance, which are early markers of chicken health concerns. Keep in mind that chickens do go through phases, such as broodiness and molting, that mimic the symptoms of some of these health concerns, so it's helpful to review these topics before deciding that they are sick (see "When a Disease Is Not a Disease" on facing page).

Behaviors That May Indicate a Health Problem

- Lack of appetite or thirst
- Watery, frothy, or especially foul-smelling poop
- Refusal to leave the coop
- Hiding
- Limping
- Listlessness
- Picking at own feathers or skin
- Cessation of laying

Physical Indicators of a Possible Health Problem

- Weight loss
- Feather loss
- Sneezing or discharge from nostrils (located on top of the beak)
- Red or discharging eyes
- Persistent swelling in chest area
- Open wounds
- Visible bugs or eggs on your chicken's skin (especially common under the wings, around the vent, or on the legs)
- Discolored comb

It is best to observe behaviors in the daytime when the chickens are moving about. However, you might find it easiest to do any close examinations at night. A chicken is much more likely to cooperate with a head-to-toe inspection when she is in her dopey nighttime state. We find a headlamp invaluable for this purpose. If you decide to proceed with any treatments or interventions, you might want to take the headlamp approach for this also. Take our word for it: it is much easier to apply Bactine to a chicken's wound when she is nearly motionless at night.

When a Disease Is Not a Disease

As we mentioned, there are some completely normal chicken processes that can seem like illnesses to a chicken keeper. Become familiar with these, and you can save yourself a panicked visit to a vet or your local chicken expert.

Molting

Molting is the term for the replacement of feathers that your hens go through every year or so. Chickens molt to some degree each fall or early winter, but many chickens who were hatched in the spring or summer will skip this process in their first year. Each chicken will have her own approach to molting. Some drop most of their feathers all at once and seem practically naked for a while. Others will go through the process gradually and look only mildly scruffy. As they start to grow

Ginger in full molt

their feathers back, you will see what at first look like porcupine quills emerging from the skin. These will grow until they eventually unfurl into feathers. While molting, your birds will stop laying.

This is a normal process and there's nothing to be done medically—you will just have to put up with fewer or no eggs at all for up to two months while the feather replacement process continues. By feeding small amounts of a high-protein snack such as shredded cheese, you can provide a nutritional boost to get the girls through this embarrassing process a little more quickly.

Broody Behavior

If your bird won't leave the nest and growls at you when you try to remove her eggs from the nest, she's not sick—she's gone "broody." Her biological clock has triggered her to stop laying and start trying to hatch her eggs. Don't bother telling her there is no rooster around and her attempts are fruitless. This a common condition of certain breeds and, beyond some folk remedies, there's little to be done besides waiting her out. As long as she occasionally ventures out for food and water, this period will end of its own accord in six to ten weeks.

During this time, you should continue to collect the eggs from under the broody hen to prevent them from rotting. That'll be one fowl in a foul mood (she is protective of her eggs), but a henpeck is rarely painful, so you can safely ignore her protests as you lift her from the nest. Your hen will often use this ousting as an opportunity to eat, drink, and take care of bathroom business.

Full Crop

If your hen has an alarmingly enlarged and firm breast, there's probably no need to worry—it's probably just full of food! Chickens will collect food in an organ called the crop located in this area. Large quantities of food may collect in the crop before moving into the rest of the digestive system. If the crop has not emptied by morning (before breakfast), you may want to consider the possibility that she is suffering from crop impaction (see below).

Loose Stools

Although diarrhea can be a sign of a serious illness (discussed shortly), it may simply be the result of something recently eaten, such as large quantities of fruit. We suggest considering this possibility before assuming that there is something more sinister going on. Try to figure out what they are eating that might be the cause, and if you think it's presenting a problem, eliminate it or control access to it.

Injuries and Physical Maladies

Sometimes a sick or injured-looking bird is just that: sick or injured. In this section, we briefly examine a number of conditions that you might encounter. We have ranked them according to how often we have observed the problem in our own flock or in those of our customers.

Pasted Vent

Basics: This is a condition that exclusively affects chicks, usually between a few days and three weeks old. The chick's vent (the outlet for both her reproductive and digestive tracts) gets plugged up with her own droppings.

Cause: Usually early stress from shipping, chilling, or overcrowding. Made more likely if the chick is eating a large quantity of nonfood items, such as litter. Often leads to (or may be directly caused by) a bacterial infection.

Symptoms: Large, easily visible collection of poop on the chick's rear. May be a large solid mass. If this has been allowed to continue for some time, the chick may get listless and lose interest in eating or drinking.

Treatment: Remove the plug immediately. See specific steps in chapter 4, page 63. We have found that persistent pasting can be a sign of a more serious internal problem. See the section on coccidiosis, page 152, for treatment guidelines.

When to Call a Vet: Consult with a vet if all or many of your batch of chicks seem to be affected. There may be an underlying illness contributing to the problem.

Prevention: Consider possible causes of stress that you could remove from the chicks' environment. Make sure it is not too drafty, cold, or crowded and that plenty of good feed and water is available. Some sources suggest that providing chick grit (small stones) in their feed early on may prevent the problem.

Predator Wounds

Basics: Predators such as raccoons, dogs, or foxes can cause significant wounds to any chicken that survives the attack. If not well cared for, these wounds can become infected and eventually become life-threatening.

Cause: You have done your darnedest to protect your chickens, but something went wrong. A motivated animal somehow got access to the birds and went about its natural business. Wounds can also be caused by other chickens, who peck out of aggression or boredom. However, we find this to be fairly rare in situations where there is adequate food and space.

Symptoms: A chicken will likely hide quite effectively after an attack if she is wounded. Once the hen is found, the wound itself will be very apparent. If there are both unwounded survivors and fatalities, be sure to make a count, then go looking for any missing birds. Look under the coop, in unused flower pots, anywhere you think a chicken could have taken shelter. Once you find your patient, she will probably be silent and acting shell-shocked. She is unlikely to run from you in this condition.

Treatment: Gently catch her and bring her to a warm, protected area for an examination. Examine her carefully, paying particular attention to the open wounds for removable foreign matter like pebbles or teeth. Because chicken skin is quite thin and delicate, it is not unusual for a large piece to be missing. As alarming as this is, a chicken can usually survive such a wound. Any deep puncture wounds into the muscle layer are more serious and prone to infection. Next, look over the hen for any broken bones. There's really not much you can do yourself to treat a hen's injuries beyond cleaning her wounds. Wash any visible soil out of the area with soapy water, then apply an antiseptic such as Betadine or hydrogen peroxide. We've also had good luck with Bactine spray, which is very easy to apply. If there is a puncture wound, consider starting the bird on an antibiotic immediately. If there is a broken leg or wing, it should be set as soon as possible. A bird of ours once broke her toe, and we did not try to set it. It's now crooked, but doesn't seem to cause her any trouble. We don't think you can get away with this if a major limb is broken.

When to Call a Vet: It is best to call the vet if you think a bone needs to be set or you want advice on what antibiotic to use. If an injury is quite likely to be fatal (major crushing injuries or puncture through the skull or abdominal cavity), a vet can help you euthanize your pet.

Prevention: Do everything you can to protect your chickens from predators. Build a secure coop and run, and lock in your flock after dark. If your chickens free range in your backyard, make sure your fences are high enough and don't have any gaps that dogs could squeeze through. Most wild predators attack at night, but neighborhood dogs can attack any time that they have access to your backyard. We once had a dog somehow wriggle free of his collar and leash, escape from his startled owner, leap over our steep front garden and a 4-foot fence, and attack the chickens in our backyard. How he even knew they were there is unclear. The dog dispatched three of our favorites and injured another before his owner was able to catch him. It was a shocking wake-up call that spurred us to increase the height of our fence to 6 feet.

Limps and Sprains

Basics: Mobility-affecting injuries are a frequent concern of our customers. However, they usually resolve or stabilize with time and rest alone.

Cause: May be anything from minor injuries to Marek's disease (discussed shortly).

Symptoms: Limping, hanging wing, or other impaired movement.

Treatment: First, rule out an open wound or abscess (see the Bumblefoot entry). If you suspect a simple strain, your bird needs rest. Some vets prescribe anti-inflammatory medicine. Provide TLC, and make sure your slow-moving bird is getting enough access to food and water. In a farm situation, a bird can die from a minor sprain simply because she cannot compete adequately for food.

In your backyard, you are in control, and your special-needs bird will likely do nicely until she recovers.

When to Call a Vet: If symptoms are getting rapidly worse or there is spreading paralysis.

Prevention: This is difficult to prevent; accidents do happen.

Crossbeak

Basics: This is a defect in which the top and bottom beak of the chicken do not line up. It can prevent the affected chicken from eating effectively.

Cause: Crossbeak is congenital. The chicken keeper can neither cause it nor stop its progression.

Symptoms: Visible "crossing" of upper and lower beaks. This often becomes more severe as the chick develops into an adult, and is accompanied by a failure to thrive or gain weight.

Treatment: In commercial farm settings, these chickens are often euthanized because their beak prevents them from eating properly, and as a result they do not grow or produce eggs like other chickens. However, many backyard chicken keepers are able to keep a crossbeak chicken alive by providing her with a way to eat even with her physical deformity. Crossbeak can be quite mild, in which case the chicken may even be able to eat like other chickens, or it might be so severe that the chicken can only survive if it is hand-fed. Other strategies include providing an extra-deep feed dish for the hen or providing moistened food that is easier for her to scoop up.

When to Call a Vet: The crossed beak generally requires regular trimming, and this is best done by a veterinarian. In mild to moderate cases, this may be all that is needed to allow a chicken to eat relatively normally. A vet can also help you euthanize your chicken if needed. We think this is the right thing to do if the chicken is simply not able to gain or keep on weight, rather than allowing it to slowly starve.

Crop Impaction and Sour Crop

Basics: Chickens have no teeth, so after they swallow their food, it moves to the crop, a holding area in the chest, before it is moved to the gizzard where it is churned about with grit (tiny pebbles), breaking it down for further digestion. Sour crop is the term used for a fungal infection of the crop. Crop impaction occurs when food is stuck in the area and unable to pass through.

Cause: Usually caused by the bird's consuming long pieces of vegetation such as grasses, which are difficult to break down and get tangled in the crop; this leads to slow or no emptying of the crop, which in turn can cause sour crop and finally starvation. On their own, chickens will usually nip off small pieces of grass, but if many long strands are lying around (such as after a mowing), chickens may ingest strands that are too long for them to handle.

Symptoms: A firm lump in the chicken's chest that seems larger than that of her counterparts, refusal to eat, listlessness, reluctance to leave the coop. Keep in mind that most chickens will have a full crop at the end of a day of eating. If you are concerned about crop impaction, check the crop the next morning before they have had a chance to eat, at which time the crop should be significantly smaller.

Treatment: The most common technique is to get the bird to swallow a small quantity of warm water or mineral oil with an eyedropper, massage the crop, and turn the bird upside down, while continuing to massage the crop in hopes that the crop will empty through the beak. (Then be aware that the contents will be stinky.) We really think that you should get some help before attempting this. You run the risk of the hen's breathing in the water, mineral oil, or crop contents, all of which would be serious bad news. However, if you think your hen's life is in danger and there's no expert available, you can try it yourself. Use the eyedropper to release droplets at the side of the bird's beak and let her swallow them herself rather than putting the dropper down her throat, lest you squirt fluid down her windpipe. Once the impaction is removed, consider giving the chicken yogurt with active cultures to clear up any fungal infection that may remain.

When to Call a Vet: A vet or local chicken expert should be able to help you empty the crop as just described. If this fails, the crop can be emptied via minor surgery—the vet will be able to determine whether this is warranted.

Prevention: A well-fed chicken who has constant access to fresh water is unlikely to eat enough grass or other roughage to cause a problem. Still, avoid giving your chickens long grass clippings, and try to mow regularly so you are not leaving long strands behind. Our mower has a mulching setting, which helps to compensate for our laziness in this area by cutting the clippings into tiny bits.

Egg Binding

Basics: An egg that is overly large or poorly formed gets stuck in the chicken's oviduct. If not attended to, this can be serious—even fatal.

Cause: Most likely to occur when a hen first starts to lay or when she is reaching the end of her reproductive years. At these times her reproductive tract is either immature or overly aged, and she is more likely to produce misshapen eggs. Also, overweight hens are far more prone to egg binding because excess fatty tissue gets in the way.

Symptoms: A hen that spends excessive time in the nesting box and appears to be uncomfortable and/or visibly straining may be egg bound. Sometimes she will not be in the nest box, but nearby and standing strangely erect as she strains to pass the egg. This should not be confused with a broody hen who is not laying at all but grumpy when disturbed at her post in the nesting box. When a hen has egg binding, you may be able to feel the egg bulging on her bottom.

Treatment: First, try simply putting the chicken in a dark quiet area with plenty of supplemental heat. Sometimes this will be enough to allow her to relax and pass the egg. If this fails, you can try to ease the egg out. First, apply a lubricant such as mineral oil to the vent and oviduct leading up to the impacted egg. Then, very gently, try to coax out the egg by massaging her lower abdomen in a front-to-back pattern. Try not to break the egg. If this happens, you will have to very carefully remove every shard from the oviduct to prevent further injury. Vets will often treat this condition by breaking the egg, but we recommend that you leave it to them unless the chicken is already too weak to move and seems to be in imminent danger.

When to Call a Vet: If the above measures fail, a vet can help the hen pass the egg, either by administering injections that stimulate stronger contractions to pass the egg or by simply breaking it.

Prevention: To make sure you don't end up with chubby chickens, feed them a well-balanced diet. Resist the temptation to give them lots of carbohydrate-rich treats. Scratch, bread, and other starchy foods should be occasional treats, not a significant portion of your flock's diet.

Prolapsed Oviduct

Basics: The bottom part of the hen's reproductive tract (oviduct) protrudes from the vent because it has been exposed to too much force.

Cause: Usually caused by a hen passing an egg that is too large for her to handle. More likely to happen when a chicken first begins to lay or is overweight or aging.

Symptoms: Your chicken is unlikely to exhibit much discomfort at first. You will see extra tissue in the area of the vent, which might look redder than usual. There may be poop clinging to the area. If the tissue looks blackened, extremely swollen, or has open wounds, it likely has been prolapsed for some time.

Treatment: Immediately separate the chicken from the flock, as they will be unable to resist pecking at the prolapsed oviduct. Carefully clean the vent with a warm damp cloth, or by running it under lukewarm water if necessary—although your chicken will protest. After this, you can try to apply some hemorrhoidal ointment such as Preparation H, as this may help shrink the blood vessels in the area and help the oviduct return to its proper place. You can try to gently push the tissue back inside, but it will likely slide right back out. If your hen is actively straining and pushing it out, you can apply a numbing ointment or gel such as Oragel. This may numb the area enough for your patient to stop feeling compelled to push. Over the next few days, keep the hen separated from the flock and keep the area clean. Over time, this problem may well resolve spontaneously.

When to Call a Vet: If the prolapse is large, quite swollen, has open wounds, has become blackened, or is getting worse rather than better over time. One of our customers—who first tried the home measures just described but lost the hen—thought that immediate vet care might have saved his bird.

Prevention: Keeping your chickens at a healthy weight will reduce the likelihood of prolapsed oviduct. Otherwise, early recognition is the best way to prevent complications and allow for natural return of the prolapse before the other chickens cause further damage.

Frostbite

Basics: Cold weather can cause frostbite to a chicken's tender and exposed areas. Usually the comb is affected, but toes can be susceptible.

Cause: Overexposure or lack of protection in the coldest weather (usually 25°F and below).

Symptoms: Discoloration or blackening of the chicken's comb, wattle, or toes.

Treatment: Usually best to leave alone. If the area is unable to recover, it may fall off. Watch for infection, in which case the damaged area may need to be trimmed off and treated as an open wound.

When to Call a Vet: If you think trimming is required, this is best done by your vet.

Prevention: This is a completely preventable problem. First, if you live in a very cold area, consider choosing breeds that do not have large combs. Those with a "walnut comb" are a good choice, for example, because the comb lies relatively flat against the chicken's head. Also, if the temperature is predicted to drop below about 25°F, consider providing supplemental heat for your coop at night. This can easily be achieved by digging out the heat lamp you used for your chicks and setting it up in the coop with an outdoor-rated extension cord.

Breast Blister

Basics: A blister that forms on the breast of the chicken. Heavy birds or lightly feathered ones are more susceptible.

Cause: Heavier birds may either perch with their breasts resting against the perch or sleep on the ground. If the perch is rough or has sharp edges, or there is inadequate soft bedding on the ground, a blister may develop.

Symptoms: A blister or scab on the chicken's breast.

Treatment: Remove the source of irritation, and this should resolve spontaneously.

When to Call a Vet: This is unlikely to be necessary.

Prevention: Provide smooth perches of adequate width. Keep soft fresh bedding on the coop floor.

Bumblefoot

Basics: A sore or abscess on the foot. Rare in hens, this usually affects the heaviest of birds—namely, roosters.

Cause: Generally begins with some kind of injury that becomes infected by bacteria, often staph.

Symptoms: You'll notice your bird limping and on closer inspection you will find a swollen area on her foot or toe. This may be red or have a black scabbed-over area.

Treatment: Antibiotics. Try Amoxicillin or Lincomycin. If you believe there is a large collection of pus under the skin, this may need to be drained. If it is scabbed over, don some disposable surgical gloves and see whether you can lift the edge of the scab to release the pus. After the wound is drained, clean it out with hydrogen peroxide and cover the wound with a sterile gauze dressing. Vet Wrap is handy, self-adhesive stuff that can go over the wound and is very tough for a chicken to remove. This will need to be changed daily.

When to Call a Vet: We do not recommend cutting open a wound yourself. If you find yourself reaching for a sharp implement, pick up the phone instead.

Prevention: Avoid possible causes of injury, especially under your coop's perch. Wire floors can cause injury when the bird hops down; rough or splintery perches are also a common cause.

Toe Balls

Basics: A collection of mud, dried chick food, or poop forms a ball at the tip of the chick or hen's toe or toenail.

Cause: Begins when a chick or hen with wet toes steps in something that sticks. This repeats over and over until a collection forms, usually appearing as a sphere of brownish stuff clinging to the end of the nail.

Symptoms: Limping may be noticeable if the collection is large enough.

Treatment: For both chicks and hens, this should be removed. Do not be tempted to just pull it off—especially if it is a chick you are treating. It is possible to pull off the whole nail and cause permanent injury. First soak the accumulation of mud to soften it, then gently remove it. If you feel like you need to use any force, then more soaking is called for.

When to Call a Vet: Unlikely to be necessary, unless you inadvertently remove a whole toenail.

Prevention: Watch those toes. If you can catch and remove any clinging material early on, the process is much easier. This is probably less of a problem where chickens are allowed to freely range and scratch.

Parasites

Parasites are among the most common health concerns that afflict chickens. Luckily, many are relatively easy to identify and treat without the assistance of a vet. We have ranked these critters from most to least commonly encountered.

Coccidiosis

Basics: Caused by an excessive proliferation of the coccidia protozoa. Usually affects young chicks, but can on occasion make adults ill. All chickens have a low level of coccidia in their guts.

Cause: Coccidia causes illness (and sometimes death) only when protozoa populations explode to a level that the bird can no longer withstand. Chicks are hatched without immunity, but gain it as they are slowly exposed over the first several weeks of life. Coccidia are spread throughout a flock through feces, and the protozoa can live in soil for years.

Symptoms: Listlessness, huddling as if cold, ruffled feathers, loose or bloody stool.

Treatment: Immediate treatment with the correct antimicrobial. We use Sulmet Drinking Water Solution, a sulfa-based antibacterial, which works very well. Isolate affected chicks or chickens if possible to limit spread.

When to Call a Vet: You can probably handle this one on your own. Just be prepared for the possibility that you might have some losses.

Prevention: Seriously consider feeding chicks medicated feed for the first twelve weeks of life. This will not kill all coccidia, but rather keeps populations at a low level that allows the chicks to build up immunity without actually becoming ill. Keep chicks' litter dry and clean—coccidia thrive in moist environments. Change food and water often. Immediately change if there is visible contamination with poop. Birds who survive illness will likely have good immunity thereafter. Proper sanitation will help prevent this infection in adults. A vaccine is available for coccidiosis, but most of the chicks available at feed stores have not been vaccinated.

Worms

Basics: Chickens pick up parasitic worms as they forage in their environment. These can cause a variety of symptoms, but are rarely fatal.

Cause: The most common types of worms that affect chickens are roundworm, tapeworm, cecal worm, and gapeworm. Most chickens up pick worms by eating an intermediate host such as snails, earthworms, slugs, or other critters.

Symptoms: Roundworms, hairworms, and tapeworms generally cause weight loss, lethargy, and a decrease in egg laying. Affected chickens may eat more to compensate, or if they are feeling quite ill they may lose interest in their food. If your chickens have gapeworms, you might notice that they are breathing with their beaks open or gasping, because this particular worm lives in the bird's windpipe and can obstruct breathing. Keep in mind, however, that open beak breathing is normal in hot weather. Cecal worms are unique because they host another organism—a protozoa called *Histomonans meleagridis*—that actually causes the illness, which can lead to liver or cecum damage. Worms may or may not be visible in your chickens' droppings.

Treatment: The most readily available wormer, piperazine, can be found in most feed stores. It is administered through the flock's drinking water. This is effective for roundworms, but not necessarily for hairworms, tapeworms, gapeworms, or cecal worms.

When to Call a Vet: If you suspect worms, you can start by worming the whole flock with piperazine. If symptoms persist, take a fecal sample to a vet for diagnosis of the responsible critter and recommendation for the correct wormer and dosage. We recommend disposing of any eggs your chickens lay for one week following worming.

Prevention: Some sources recommend worming twice a year as a matter of course. We feel this is excessive for most small backyard flocks, unless worms have been an established problem in the past. In fact, your chickens may well have a small number of worms on board at any time, but will not have any symptoms at all unless the number of worms is overloading their systems. If you do choose to worm your chickens preventatively, do so when your chickens are molting. They will not be laying as much during this time, and you will waste fewer eggs.

Lice

Basics: Lice are among the most visible parasites that could afflict your birds. They generally eat only shedding skin, scabs, and feathers, but they can cause a significant amount of irritation in the process.

Cause: Lice are introduced to a flock if another critter with lice lives in their environment. Often this occurs when an affected bird is purchased and added to the flock.

Symptoms: Visible signs of itching and irritation, such as compulsive and overzealous grooming and self-picking. Grey or tan flat-bodied bugs may be visible, especially under the wings and around vent. Under close inspection, you might also see collections of gray eggs clustered around the feather shaft. The feathers may look shredded or less glossy from louse damage.

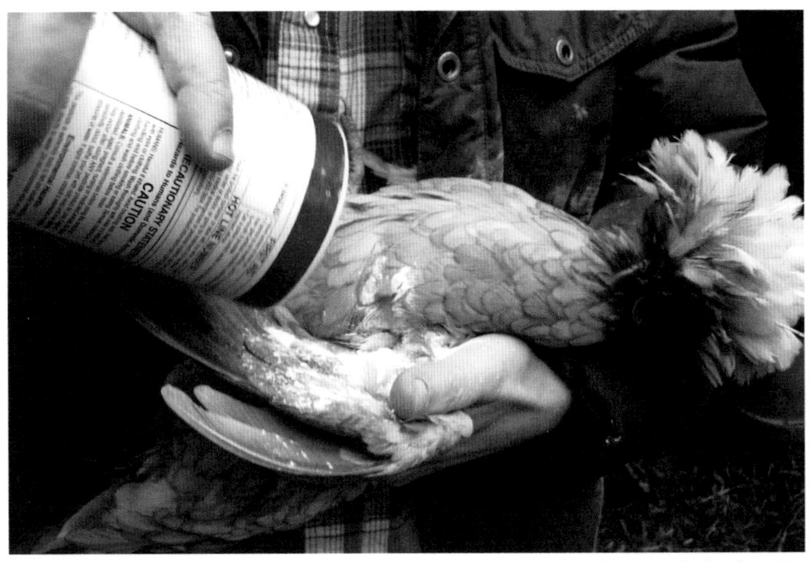

Dusting a chicken for mites

Treatment: Powders are available for applying directly to your affected chicken's feathers and skin. Follow the instructions carefully. Any visible collections of eggs should be scraped off (wear gloves for this) and disposed of in the toilet or burned.

When to Call a Vet: If you have an infestation that is affecting a batch of chicks, as this can be fatal for very young birds.

Prevention: Check the feathers and vent of any bird you are considering purchasing for signs of lice infestation. Provide your flock with facilities for dust bathing or dust them periodically with poultry dust (a powder usually containing mild insecticide synthesized from a naturally occurring substance found in the chrysanthemum plant). Before dusting your coop, clean out and replace the bedding material inside the coop.

Northern Fowl Mite

Basics: These are tiny little bugs that live on the hen and cause chicken misery.

Cause: Your flock can be exposed by wild birds or by a new addition to the flock.

Symptoms: A severe infestation can cause anemia. You may notice that your chicken looks paler, especially around her comb; her egg laying may drop off;

and she will seem itchy, irritable, and uncomfortable. The mite eggs are generally more easily seen than the mite itself, and are most likely to be found on her back or around her vent.

Treatment: Poultry dust is the solution. Apply it as instructed. It will not kill the eggs, however, so reapplication is necessary to kill the next generation of mites before they lay more eggs.

When to Call a Vet: Unlikely to be necessary. Consider calling if you have followed all instructions and symptoms of illness persist.

Prevention: Provide dust-bathing facilities or use poultry dust periodically (see the lice entry).

Red Mite

Basics: These nasty little blood drinkers lurk in the darkest, most hidden areas of the coop and come out at night to feed on your hens.

Cause: Same as for northern fowl mites, but generally lives in the coop versus on the birds.

Symptoms: Your bird will be very uncomfortable from the bites she has received; she will scratch and groom without rest, sometimes even pulling out her own feathers in desperation. Signs of anemia, such as paleness and cessation of laying, may also be evident.

Treatment: Because these mites do not live on the hen full-time, you need to treat their hiding holes as well as the hens themselves. It is easiest to spot them at night, when they emerge. Treating the coop should involve vigorous washing of the coop, ideally with a pressure washer, and then spraying down with a disinfectant. Pay special attention to gaps and hidden areas such as corners and under the perch.

When to Call a Vet: If anemia seems severe or life-threatening, or you can't seem to clear an infestation.

Prevention: Cover the coop floor with clean bedding material. Every time you perform a thorough coop cleaning, disinfect it as just described. Once the coop is dried, sprinkle a little food-grade diatomaceous earth (or poultry dust) in the corners of your coop before replacing the litter. This may prevent many types of creepy-crawlies from getting a foothold in your coop.

Scaly Leg Mite

Basics: Mites that burrow under the scales on a hen's legs, causing visible distress to the bird.

Cause: Exposure to another affected hen. More likely to become a problem if your chickens live in damp litter or are feather-footed.

Symptoms: The scales of the legs will become crusty and deformed, often lifting up to form little homes for these unpleasant critters.

Treatment: Try to catch this as soon as possible; the more established the mite, the more hiding places it creates in the scales, and the more difficult it becomes to eradicate. The age-old remedy is to coat the legs with Vaseline daily, which will eventually smother the mites. This is a nontoxic approach, but it will be slippery and messy, and your hen will object vigorously. There are strong veterinary antipest sprays that may be used, but any eggs laid should be discarded for at least a week after discontinuing use of such a product. Do not try to pry up lifted scales, as this will be excruciatingly painful for your bird. The scales will likely remain lifted even after the mites have been successfully controlled. Normal scales will reappear the next time the hen molts.

When to Call a Vet: If home treatment is not resolving the problem within a month or so.

Prevention: Do not purchase a hen with crusty, scaly legs. Start with clean bedding material inside the coop, and keep the coop clean and dry. Keep a close eye on any feather-footed chickens; not only are they more susceptible, but also their feathers may prevent you from seeing the problem early on unless you are watching for it.

Ticks or Chiggers

Basics: More common in warmer areas. Ticks are brief visitors who take their meals to go. Young chiggers will cling and feed for a few days, then hop off, leaving little red dots of extremely irritated skin. They generally feed at night, then hide during the day.

Cause: Chickens pick them up from their surroundings.

Symptoms: In the case of ticks, you are looking for symptoms caused by excessive blood loss. Paleness, a drop in egg production, and weakness are all common. These same symptoms apply to chiggers, but if they are responsible you will notice your chickens looking seriously uncomfortable and itchy. If you have ever suffered from a chigger yourself, you will empathize.

Treatment: Treating the coop is the best way to keep ticks or chiggers at bay if you suspect that they are your culprit. As with mites, vigorously wash the coop, ideally with a pressure washer, and then spray down with a disinfectant. Pay special attention to gaps and hidden areas such as corners and under the perch.

When to Call a Vet: A vet will best be able to recommend the products for treating the coop that will be most successful in your area.

Prevention: If practical, remove tall grasses and other tick and chigger havens from your chickens' environment.

Illness and Disease

When most people think of sickness, they think of bacteria, viruses, and the like. Because there are so many different "bugs" that can affect chickens, we have put the three most common types in this section and grouped the rest in the section that follows it.

Upper Respiratory Infections

Basics: There are many potential causes of coldlike symptoms or breathing difficulties in chickens—so many, in fact, that they are very difficult to differentiate without the help of a vet.

Causes: Any one of numerous bacterial, viral, or even fungal organisms. Nevertheless, if your chicken looks like she has a cold, you should follow the same instructions.

Symptoms: Coughing, sniffling, and sneezing. In severe cases, your bird might be struggling to breathe, or you might be able to hear a gurgling or crackling sound as she inhales.

Treatment: Immediately remove the symptomatic chicken from the flock to prevent spread of the illness. We recommend providing supportive care in the form of heat, excellent cleanliness, plenty of fresh water, and extra protein in the diet. A product called Vet RX may help clear nasal passages. If the hen is not improving over the next few days, and particularly if she is looking quite ill, it is time to add antibiotics. Erythromycin or Sulmet are good first choices. Your chicken may be suffering from a bacterial illness, or if the original cause of illness was viral or fungal she may be experiencing a secondary infection.

When to Call a Vet: If the illness seems severe and is spreading rapidly through your flock, and the treatments just suggested are not helping, it is time to ask a vet to help you figure out exactly what is causing the outbreak. This is important because certain illnesses can be quite contagious, and you want to know how to appropriately decontaminate your setup before adding more birds. Use caution in such circumstances before visiting other chicken keepers (or feed stores), lest you spread the illness via contamination on your shoes. Vaccines are available for many chicken illnesses and may be a very good idea indeed for the health of your future chickens once you have a history of exposure.

Prevention: Avoid visiting the area around the coops of folks with ill birds. Buy only very healthy-looking chickens to add to your flock. Ask the person who raised them whether there have been any past incidents of illness. If there are any questionable issues, either do not buy the birds or consider a period of quarantine before introducing them to the rest of your flock.

Marek's Disease

Basics: An extremely common disease for chickens that is frequently fatal.

Cause: A form of herpes virus that can cause tumors of the nerves and organs of the unfortunate chicken who is exposed. Infectious particles are carried on feather dander and so are spread about easily in the wind. If chickens inhale the infected dander, they can acquire the virus. Unfortunately, it is also an extremely hardy virus; it can live in the environment without a host for a year or more.

Symptoms: Paralysis of varying degree, often on limbs on one side of the body or the other. Enlarging of the iris and lack of definition between iris and pupil, also often on one side. Dilation of the feather follicles, trouble breathing (because of paralysis of the lungs).

Treatment: None known.

When to Call a Vet: If you want help to euthanize a suffering chick or chicken.

Prevention: Make sure that the chicks you purchase are vaccinated against Marek's. This is usually done at one day old right before shipping, or even while they are still in the egg! If you purchase grown hens, ask whether they were vaccinated, and check their eyes. Some chickens do survive Marek's but continue their lives as carriers of the disease. These will often have a fuzzy border between pupil and iris. Unfortunately, vaccination does not provide 100 percent protection, so it is possible to lose a chicken to Marek's anyway. It is the single best step you can take, however, so insist on vaccination.

Omphalitis (Mushy Chick Disease)

Basics: Infection of the umbilical area from unclean conditions at time of hatching—uncommon in good commercial hatcheries.

Cause: Usually a bacterial infection of the yolk sac, which is absorbed into the chick's abdomen at hatching. Usually occurs in situations where eggs are hatched in contaminated or overly humid incubators. Attempted hatching of cracked eggs can also be responsible. It is not contagious from chick to chick, but chicks from the same hatching are likely to be affected.

Symptoms: Sometime between hatching and fourteen days old, affected chicks will suddenly become lethargic and behave as though they are chilled.

Close examination will show a swollen umbilicus (located farther up on abdomen from the vent). Often symptoms do not occur until shortly before death.

Treatment: We are aware of no known effective treatment. Antibiotics have not been shown to increase survival rates significantly.

When to Call a Vet: Not necessary if you are fairly sure you are dealing with mushy chick disease.

Prevention: Purchase chicks from reputable hatcheries. If you hatch your own chicks, use only clean, uncracked eggs, and make sure your incubator has been disinfected properly between hatchings.

Other, Less Common Conditions

These are illnesses that you would likely need a vet to diagnose definitively. Nevertheless, it may be helpful to know what is out there. Because you are unlikely to encounter these, and even less likely to be able to treat them without help, we've kept the information basic.

Infectious Coryza

Your chicken will look as though it has a head cold. It will sneeze, breathe loudly, and have plenty of nasal discharge. The face may be swollen, and the eyes will likely be shut from sticky discharge. The organisms responsible are bacteria, so antibiotics are generally effective. Erythromycin or Sulmet are good first choices. Separate an ill chicken from the flock to prevent spread. There is a vaccine, but you should consider it only if you have experienced coryza in your flock before.

Avian Mycoplasmosis

This is caused by a fungal toxicity that produces upper respiratory symptoms similar to those of other poultry illnesses. A common cause of chronic respiratory disease. Treatment with antibiotics is helpful to prevent or treat a secondary bacterial infection. Now largely eradicated from commercial flocks in the United States, it is still found occasionally in backyard flocks.

Colibacillosis

This digestive tract infection is caused by *E. coli* bacteria, commonly found in the environment, but chickens tend to get sick only if they have excessive exposure to unclean litter. The ill chicken will have terrible diarrhea, and her feathers will often be matted around the vent. She will lose weight and look miserable. Erythromycin is effective. Treat only affected hens, and clean out and thoroughly disinfect your coop with a diluted bleach solution. Make sure the coop is entirely dry before allowing the chickens back in.

Fowl Cholera

A bacteria called *Pasteurella multocida* causes either acute infection, which usually leads to death, or chronic infection. Infected birds look very sick before they pass away. They will not eat, and they may have a mucus discharge from their mouths, diarrhea, and rapid breathing. A chronic infection will produce swollen wattles, joints, and eyes. This disease can be carried by rodents— a great motivation for keeping those critters under control. Antibiotics are effective, and the coop can be cleaned with disinfectants. It's fortunate that this is a rare poultry disease.

Fowl Pox

This is caused by a virus that may cause sores or scabbed areas in the mouth or feather-free areas. Insects are responsible for spreading it from one bird to another. There is no specific treatment, but you can provide supportive care to help your flock through. Provide heat if it is cold, and be extra vigilant about providing clean food and water. A vaccine is available.

Newcastle Disease

This nasty viral affliction has no known cure once birds are infected. A vaccine is available, however. Symptoms vary from coldlike to paralysis. Mortality depends on the strain that your flock is exposed to. The disease generally passes from one afflicted flock to another on the bottom of shoes or other contaminated items—

a very good motivator for taking the general precautions mentioned earlier when returning home from other farms or flocks.

Aspergillosis

The disease caused by this fungal organism is also called brooder pneumonia because the spores are generally breathed in by chicks—from contaminated litter—shortly after hatching. It's rare in small flocks that are kept in clean conditions. An affected chick or chicken will likely have trouble breathing and rarely survives if symptoms such as this arise. If you suspect aspergillosis, contact your vet for confirmation and to learn how to thoroughly sanitize your chicken setup.

Avian Flu

There are many strains of avian influenza, but a particularly aggressive strain of the virus called H5N1 has received much media attention in the past several years. As of this writing, it has not yet been found in the United States. The danger is that the virus could mutate and become contagious from human to human. As this concern has not yet been borne out, we recommend not spending much time worrying about it. Any outbreak that occurs is likely to be subject to extremely aggressive infection control measures. If you are worried anyway, then do what you can to keep your chickens from socializing with wild birds. The easiest way to do this is to confine them to a coop and run with a roof.

• • •

We hope that this chapter has been helpful, but not overly intimidating. Keep in mind that we, as chicken keepers, have personally had to deal with only physical injuries to our chickens. This is fairly representative of the average chicken keeper's experience. We doubt you will ever encounter avian mycoplasmosis or scaley leg mite, but at least you will not be helpless if you do!

FROM
Coop
TO KITCHEN
in Zero Miles!
REDUCE YOUR CARBON FOOTPRINT
BY RAISING YOUR OWN CHICKENS

8

EGGS: HOW TO MAKE THEM AND EAT THEM

CHICKENS MAKE EXCELLENT PETS in their own right, but for most of us the perfectly fresh, delicious, and nutritious eggs, just steps from our back door, are the most compelling aspect of chicken keeping. You will soon discover that an egg laid by one of your hens is a world apart from the anemic, runny things you have become accustomed to from the supermarket. Your backyard eggs will be far and away more delicious and nutritious than even the most expensive free-range, vegetarian-fed, hand-coddled dozen you can buy at your local specialty store.

In this chapter we explore the wonderful world of eggs, from their structure and function to the practical knowledge you'll need to keep your flock happily laying. We share a few of our favorite recipes for your backyard jewels that range from perfected versions of the familiar basics to a few exotic treats and recipes from around the planet. We also discuss food safety, even revealing the shocking truth about the expiration date on store-bought egg cartons. Last, we'll offer a roundup of the latest research and reporting on the ever-changing scientific opinion about eggs and your health.

ALL THEY'RE CRACKED UP TO BE

An egg, as we all know, is a reproductive structure designed to feed and protect an embryo outside of a bird's body. In the wild, birds will lay their eggs during one or more favorable times of year when food and other conditions are optimal. After laying a small collection of eggs, mother (and sometimes father) birds will then spend long periods of time brooding their clutch, essentially by sitting on them. Their body heat is transferred to the eggs from their skin and retained by downy feathers, allowing the embryo to develop rapidly inside the protective shell. After a while, the chick pecks its way out of the shell and emerges.

Chickens reproduce in the same way, but unlike wild birds, domesticated laying hens will lay during most of the year and continue to lay even as the eggs fill their nests to overflowing (to a point). They do this because humans have selected for this trait through many generations of preferring birds that lay profusely over those that exhibit strong brooding instincts. Remarkably, most types of chickens bred for egg production will lay about two eggs in a three-day period. However, an egg is not made in a single day; indeed, its development is quite an involved process. At any given time a hen will have three or more eggs in the works, at various stages of completeness, along her oviduct.

One of the most common questions we field is whether a hen will be able to lay, or lay as well, without a rooster. In urban settings, this is an especially pressing concern because roosters are generally prohibited or at least discouraged. In fact, even if you are allowed to keep a rooster in your city, we would encourage you to resist as a courtesy to your neighbors. They are truly beautiful creatures, but they are loud, and they crow frequently, not just in the morning, but throughout the day and even at night if disturbed.

That said, you will be happy to hear that the presence or absence of a rooster will have no impact on your hens' egg productivity. All a rooster does is fertilize an egg inside the body of a hen before it is deposited in the nest. The rest of the process continues unhindered without him. You only really need a rooster if you are hoping to hatch chicks from your eggs.

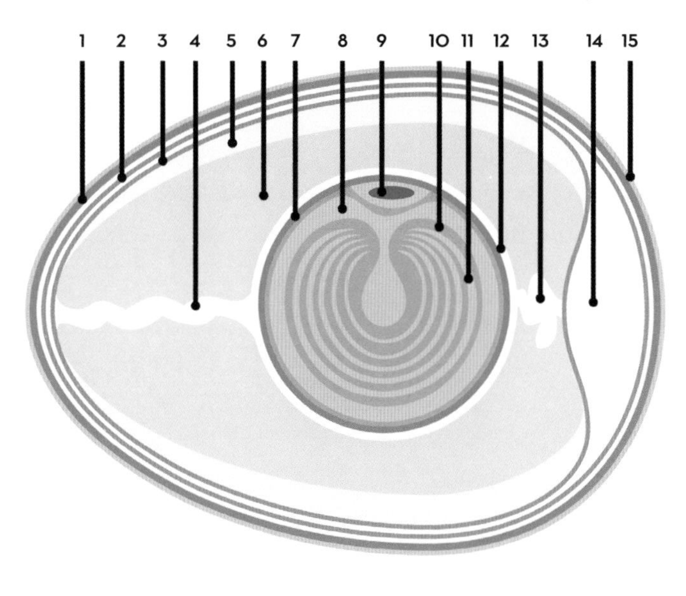

1. Eggshell
2. Outer membrane
3. Inner membrane
4. Chalaza
5. Exterior albumen (outer thin albumen)
6. Middle albumen (inner thick albumen)
7. Vitelline membrane
8. Nucleus of pander
9. Blastoderm
10. Yellow yolk
11. White yolk
12. Internal albumen
13. Chalaza
14. Air cell
15. Outer cuticle layer

Eggs are fascinating structures. Although you may not need all of the detail provided in the diagram, it is useful to know some basics. The first thing to note is that in addition to the protective shell (1) that we are all familiar with, there is an outer cuticle layer (15). This thin coating helps protect the egg from drying out and otherwise spoiling. If you wash your eggs after collection, you will remove this layer, potentially shortening their shelf life. In a related way, the air cell (14) is an important indicator of freshness. As the egg ages, this area will gradually increase in air volume. As the volume of air increases, the egg will eventually become buoyant in water, providing a useful indicator of spoilage. In the section on egg preservation we will discuss cleaning and storage considerations in more detail.

Next, we'd like to point out the chalaza (4 and 13). These structures help to suspend the yolk (7–12) safely in the middle of the egg's middle albumen (6) or "white." Although strong and flexible, these fibers have their limits and can give out if the egg is roughly handled. For this reason, and others, it's important to handle your hens' eggs carefully, or the yolk may be broken within the intact shell. Broken or malformed yolks should be more of an issue with store-bought eggs that have traveled long distances than your homegrown beauties. You will probably notice the chalaza more in your hens' eggs than in others—this is a sign of their freshness!

Lastly, let's turn our attention to the yolk (7–12). The yolk is the most nutritious part of an egg, and for good reason: the yolk is what feeds the developing embryo until and even beyond hatching. The first thing everyone notices about their homegrown eggs is that the yolk is usually much firmer and deeper yellow (or even orange) than what they're used to. This is due primarily to the diet your birds enjoy. Not only can you afford to feed your birds better than most farmers with huge flocks, but your hens will have at least some access to green plants. It's the chlorophyll in the green plants that makes your yolks so special. More on this in the nutrition section that follows.

LAY WHILE THE LAYIN' IS GOOD

A hen's ability to lay will be determined by a combination of environmental factors and her genetic makeup. A chicken who was bred for the beauty of her feathers, like a Crested Polish, will not be able to compete with a White Leghorn when it comes to cranking out the eggs. This is simply because her ancestors were repeatedly selected for their appearance rather than their laying abilities. A typical hen of an egg-laying breed will lay between 150 and 300 eggs a year during her peak laying years, the first through second. After that, production will taper off by about 20 percent a year, until by the sixth year (if she lives that long) any eggs at all will be a bonus.

You will have a large role to play in providing good conditions for optimal laying. A chicken's ability to crank out the eggs is mightily influenced by her environment. A number of environmental factors combine to influence laying, including temperature, day length, feed quality, and stress. When these elements combine within optimal ranges, egg production can be pleasingly predictable and productive. When any one of these factors goes awry, production falls off noticeably. It's for this reason, not necessarily a disregard for their chickens' well-being, that commercial egg farmers keep their birds in environmentally controlled barns.

Light

Your chickens will be at their laying best during the spring and summer months when the days are longer. Conversely, as the days become shorter in the fall and winter, you will see your egg supply dwindle. Light (and to a lesser extent heat) is the key part of the equation here, and it is something that you can influence

Multicolored eggs

easily. Increasing the hours of light makes the chickens think the days are longer, and their egg production picks up. To do this, we hang the same lamp we use to brood our chicks out in the coop and set a simple appliance timer to run from 4:00 A.M. to 9:00 A.M., providing an additional five hours of light per day through the winter months.

We simply run out an extension cord to the coop and hook up the light at the highest point of the interior, facing the roosting bars. Then we set the light cycle with the timer. It's very important to make sure that the light fixture is secure inside the coop so it cannot fall, as the hot bulb poses a potential fire risk. Although we have had success using the red bulb recommended for chicks, this technique works better with a white or full-spectrum bulb.

If you are providing supplemental light to increase egg production, we recommend setting it up to illuminate the coop in the early morning hours while your birds are still cooped up. This not only renders them a captive audience to the light, but prevents bedtime confusion. This past fall we heard from many perplexed customers who reported that their chickens were suddenly failing to return to their coop at night. When we asked, these customers invariably had set up a light to extend the chickens' light exposure, but mistakenly included an evening cycle. This won't work well because it interferes with the chickens' instinct to return to the safety of their dim coop as the sun sets. They will often cluster right outside of the coop's door into the night, quietly clucking in confusion, leaving them vulnerable to weather and predators.

Molting

Another factor that affects egg production in the fall is molting. This is a normal, usually annual process in which a chicken sheds her older feathers and replaces them with shiny new plumage. From an egg-laying point of view, you need to know that the energy required for this will interrupt laying altogether for a month or more. For this reason, most commercial egg farmers "retire" their flocks as they begin to molt and replace them with a new flock. At home, you will want to help your tatty-feathered friends get through the process quickly by offering a higher-protein diet for a few weeks. See chapter 7 for detailed recommendations.

Broodiness

As we've said, the hens of some breeds have a tendency to "go broody," at which time you will notice that your normally active and friendly chicken is now spending all her time in the nesting box and pecks grumpily at you when you try to collect the eggs. She may even make a strange sound, essentially the chicken equivalent of a growl. No, she has not lost her senses, she is trying to hatch an egg. Unfortunately, hens who are prone to this behavior will do so whether or not there is a rooster around. She does not know that her efforts are in vain.

You can try to "break" her broodiness, which involves confining her away from the nest for a few days. But for most urban chicken keepers the hassle of building another safe place for her to be for those days is prohibitive, so we generally suggest just waiting it out. While a chicken is broody, she will generally leave the nest only once or twice daily to eat, drink, and take care of bathroom business. Keep an eye on her, and make sure she is doing so. We've heard that some broody hens are so dedicated that they don't leave the nest at all of their own will and run the risk of starving. However, if you pull your chicken off her nest and plop her in front of some feed, she will eat. The most reliable way to avoid broodiness completely is to choose breeds in whom this trait is essentially bred out. A leghorn that goes broody is a rare bird indeed (see "Broody Behavior," page 142, for more information).

Nutrition for Laying Hens

What you feed your hens will play a large part in determining how well they lay and the even the nutritional quality of the eggs themselves. As we discussed in a previous chapter, chickens who eat a diverse diet based on quality chicken feed that also includes greens, insects, and other foraged items will be receiving optimal nutrition. Two elements in particular have the most influence on your birds' ability to lay well: calcium and protein.

Soft-Shelled Eggs

Occasionally, a hen may lay an egg with a soft shell, or even no shell at all. If this occurs when a young chicken is just getting started laying, or after significant changes in the weather or season, or after molting or broodiness has temporarily stopped production, it is not cause for alarm. In these situations, or most others for that matter, the main thing you can do for her is to boost the amount of calcium and protein she is getting. This can be accomplished by providing oyster shell and cutting back on high-carb treats. It is also possible that your hen is exhibiting a nutrient deficiency. Try adding a vitamin powder to your flock's water for a few weeks. This may resolve the problem.

Calcium is an essential mineral for the production of eggshells and other bodily functions. Any fresh, carefully formulated layer feed should contain sufficient calcium to support laying needs for most times of the year. Calcium is usually provided in the form of limestone, though almost any relatively pure source of calcium will do. During seasons conducive to heavy laying, you will need to supplement with additional calcium to support the formation of all of those shells. As mentioned previously, we do not recommend the common practice of feeding back eggshells to your hens, as this can lead to the dreaded (though uncommon) problem of hens eating their own eggs (which may also be a sign of protein deficiency). If for some reason you must feed eggshells, they should be toasted and crushed to reduce their resemblance to an egg. To keep things safe and simple, we strongly recommend using only oyster shell as a calcium supplement. In addition to bearing no resemblance to a chicken egg, we like to use oyster shell because of its micronutrient content, ease of storage, and low cost.

Protein is also an essential component of a chicken's diet. Its quality and abundance will determine, in large part, both the size and the quality of the eggs produced by your flock. Layer feed will contain protein from one or more of the following sources: soy meal, fish meal, meat by-products (though this has been largely phased out of commercial layer feed), plant by-products, or other regionally abundant sources. Whatever the source, most layer feed contains between 16 and 18 percent protein. Again, this should be sufficient for moderate egg laying, but you should consider supplementation during periods of heavy laying, molting, and stress. Protein supplements include low-sodium shredded cheese (our favorite), cottage cheese, milk, canned tuna, and cooked legumes. Some people feed small bits of lean meat to their hens, but others feel that terrestrial meat sources should be avoided. Most experts now agree that feeding chicken to chickens is a dangerous practice. We just think it's creepy and wrong.

Retirement

As mentioned at the top of this section, egg-laying hens will produce eggs abundantly during the first few years of their lives and then taper off to almost nothing as they age. There is much debate among small flock keepers about what to do with older, less productive birds. In commercial settings, they are seldom allowed to live past their second year, but are slaughtered and their meat used in pet foods or other processed meats (chickens bred for meat are of an entirely different breed and have an even shorter life). At home, older birds usually maintain their status as pets regardless of their egg-laying shortcomings and are kept until they die a natural death.

Others folks, usually citing feed costs, will want to periodically replace their older birds with younger, more productive ones. If you are lucky enough to have a livestock sanctuary in your area that will accept your retired hens, this is a great option that we recommend heartily. If you have not become too attached to your hens or otherwise have a strong stomach and a deep practical streak, you can do as some of our customers do and "harvest" them yourself. The strongly flavored meat will bear little resemblance to what you are used to, so it's best used in stews and other recipes that will soften its tough fibers.

EGG SAFETY AND STORAGE

Crack one of your home-grown, still warm eggs into a frying pan side by side with a store-bought egg and you'll see that there's really no comparison. Thanks to your efforts for your hens in the production of these beauties, you can be fairly confident that the eggs they lay will be of remarkable quality and freshness. To ensure that they remain that way, you will want to take a few simple steps.

First, we strongly recommend collecting the eggs at least once a day. We check twice, because we're out at the coop twice a day anyway, letting the chickens in and out. This way, we get the eggs of any early birds right away, and those of the procrastinators before nightfall. The eggs go right into the fridge, which keeps them fresher longer. An egg that is refrigerated immediately after laying is about as fresh a week later as one that was left at room temperature for one day. Although we recognize that eggs are often left at room temperature in Europe and other countries, we suggest cooling them if you want to keep them for more than a week or two.

Collecting eggs also helps to protect them from breakage by clumsy hens. Chickens have the peculiar habit of wanting to lay their egg where everyone else is laying. Even if you have several nest boxes, you will often find 90 percent of your eggs in one box. As a chicken settles down to lay, she will shift around the eggs that are already present, which puts them at risk for breakage. The loss of a single egg once in a while is not a disaster, but chickens are curious and will taste anything that looks slightly like food, and once a chicken gets a taste for egg, it is very hard to break her of it. Prevention is definitely the way to go here. Providing plenty of soft bedding will also help prevent breakage (see page 176 for suggestions for nesting box bedding). It's a rare chicken that will look at an intact egg and think of it as food unless she has a reason for it. Perhaps she has tasted an egg that broke accidentally, or maybe she was given broken eggshells by her owner to supplement her calcium intake.

Backyard eggs tend to be consumed sooner than any commercially available egg, as those must be collected, sent to independent processing firms, sorted, washed, packaged, shipped, placed in grocery coolers, then eventually purchased and brought home—a process that can take weeks. The dating on a carton of eggs provides clues—but is coded. It will usually have a sell-by date, but it also has the date that the eggs were packaged as a "Julian date," in which January 1 is 1 and December 31 is 365. Because our chickens are currently molting, we happen to have a carton of fancy free-range store-bought eggs in our fridge. On this particular carton, the Julian date is 316, which means they were packaged on November 12th, with a sell-by date of December 24. That seems like long enough, but keep in mind that we don't know when these eggs were actually *laid*, only when they were *packaged*.

Eggs keep for the longest time at 45°F–50°F and at 75 percent humidity. Conveniently, these happen to be the conditions of an egg tray in the door of a refrigerator. They can be stored in cooler temperatures, but eggs will break below about 29°F. Well-stored eggs will keep for about two months.

We recommend that you store all eggs in cardboard cartons even when in the fridge door. Cartons will help keep them safe from jostling and will moderate the humidity. It's purportedly better for the eggs to be stored with the pointy end facing downward so that the yolk stays suspended better in the center of the shell.

You may find that your midsummer abundance of eggs has become a distant memory by midwinter. For this reason, it's always a good idea to stash a few eggs for later use. If you wish to store eggs for very long periods of time, the easiest method is freezing. A whole egg may explode when frozen, so it's best to crack the egg and lightly beat it before freezing. The cells of a large ice cube tray will each accommodate the contents of a single egg. They can be frozen in the tray, then transferred to freezer bags and labeled with the date.

Egg Hiding

All of our dutiful egg collection has had one unintended consequence. Chickens have their own agenda when it comes to egg laying, and it's not to have you collect them. They are trying to lay enough eggs, settle into the nest, and brood them until they hatch. Frequent egg collection helps to thwart this instinct, but it sometimes leads the hens to seek a safer place to lay where the eggs don't keep disappearing. You may think that your hens have mysteriously stopped laying, only to stumble upon a cache of a dozen or more eggs that they have been hiding from you. These eggs are usually dirty and have had no refrigeration, but we'll show you how to use the float test to check them for freshness (see below).

Although we can't say that we have any research to support this, we have found that leaving one egg (or a good fake egg) in the nest when we collect the rest seems to be enough to keep the chickens laying where we want them to. Robert thinks that this works because chickens don't count very well.

Broken Eggs

You may find that your hens' eggs are consistently cracked or completely broken. This is undesirable for several reasons. First, you will probably not feel safe eating a cracked egg—the inside has become slightly exposed to the outside world and may have become contaminated. The other major problem is, again, that your hens may be tempted to sample the tasty white or yolk, leading to egg cannibalism.

To prevent egg breakage, feed a proper diet and make sure that the nest box is lined with a soft and sanitary material like hay (not straw). You should also consider using a plastic, artificial-turf-like nest liner made for this purpose. Whatever you choose, make sure that it will not harbor insects and is easy to wash occasionally. We use nest liners with fresh pine shavings on top, a combination that has worked beautifully.

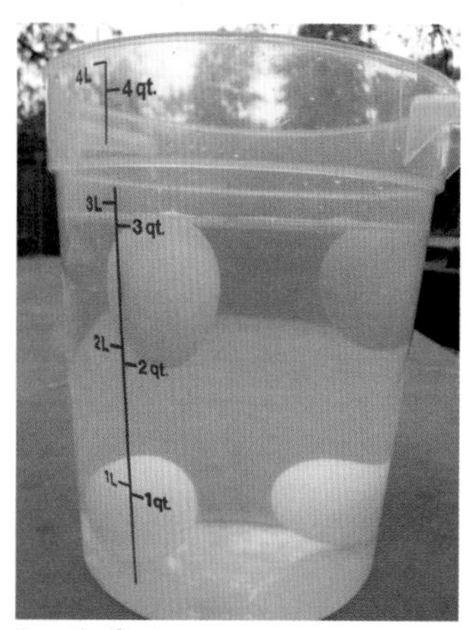

Toss the floaters

The Float Test

An eggshell is covered with tiny pores that release moisture and allow air to seep in as the egg ages. As this process continues, a small air pocket inside the shell will grow. Think of the last hard-boiled egg you ate. Remember that peculiar flat area on one end? That reflects the air pocket that was present when it was cooked. The flatter the area, the larger the pocket, and the older the egg was. This air pocket provides a handy tool for determining the age, and thus the edibility, of an egg. Immerse the questionable egg into a bath of cool water. If it sits horizontally on the bottom of the dish, it is quite fresh and you have several weeks to eat it. If it stands up on one end but still touches the bottom of the container, it has some collected air and should be eaten in the next few days. If it floats in the middle of the water volume or on the surface, it has aged and should be disposed of. We use this little trick whenever we find eggs that one of our chickens has hidden. We have no idea how long she has been up to this sneaky business, and we don't want to break a substandard egg into a recipe that is under way.

Dirty Eggs

As we mentioned when explaining the egg diagram, the outer cuticle provides a layer of protection that helps keep the egg from drying out. If you come across an egg that has been soiled, usually occurring when a hen has been sleeping in the nest, the temptation is to wash it immediately. Unfortunately, washing the egg will remove the cuticle, hastening spoilage. The tiny pores in the eggshell will allow moisture to escape more quickly without the cuticle, and theoretically they can also allow any bacteria that remain on the surface of the shell to enter the egg. You have three choices at this point: leave the dirt on and wash the egg immediately prior to using it, wash it and disinfect it, or wash it and replace the protective layer before storage. We prefer to wash and then apply a light coating of mineral oil. Eggs treated in this way have a deep, rich glow. However, we cannot guarantee that this will eliminate all bacteria from the surface of the egg. Always wash the egg in water that is warmer than the egg, because if the water is cooler, it could theoretically cause the egg contents to contract and suck any surface bacteria into the egg once the cuticle is washed off.

We certainly recommend washing any eggs you plan to give away, as other folks might be a little put off by a muddy or otherwise soiled egg. If you plan to sell the egg, there are regulations that you must follow. These vary depending on where you live, so look into food handling regulations in your area. This generally involves washing the eggs and then applying a sanitizer such as bleach.

Commercially available eggs are always washed and sanitized. See our Resources list for links that detail methods for egg washing and sanitizing.

Dangerous Eggs

Since we were children we have been warned not to eat raw eggs because of salmonella. How much of a concern this in a backyard setting is unclear. According to the American Egg Board, one in twenty thousand eggs actually has detectable salmonella bacteria inside it. The folks at the Board figure this means that you will run across such an egg every eighty-four years or so. If this egg is properly cooked before it is eaten, you will never have a moment's discomfort. Refrigeration and freshness also prevent growth of the bacteria, so a backyard egg, properly gathered and chilled, should logically have very low quantities of bacteria even if contaminated. Still, make sure to discard or immediately use any cracked eggs, as these have lost the natural protection provided by the shell.

We want backyard chicken owners to keep in mind that, like all creatures that provide food, your chickens should be protected from toxic substances in their environment. Theoretically, these substances could end up in the eggs you will be eating. Use common sense: If you live on top of a former toxic dump site, think twice about keeping chickens. Keep your chickens away from that leaky old underground oil tank. Avoid using potentially toxic pesticides or fertilizers in areas where your chickens roam. A less obvious source of possible contamination is lead paint from your older house, especially if it has been recently scraped and repainted. We think it makes sense that if you have flaking or peeling lead paint, your chickens should be excluded from areas where they can easily peck at it. Usually, this type of contamination is most concentrated within 3 feet of the house.

Note: The Centers for Disease Control recommend that you avoid consumption of soft or uncooked eggs. While we feel that your eggs should be safer than commercially produced ones, we want to pass this warning on. This is especially a concern for folks cooking for the very young, very old, or those with a compromised immune system.

NUTRITIONAL VALUE OF HOMEGROWN EGGS

It seems that eggs are in and out of nutritional fashion every decade or so. Currently, the American Heart Association no longer recommends specific limits on how many egg yolks to eat per week. This recommendation reflects results from long-term, large-scale public health studies that have failed to find a correlation between increased egg consumption and heart disease.

FUN FACT
All of the egg's vitamins A, D, and E can be found in its yolk.

For years we have heard rumors that small farm and backyard-raised eggs were better for your health than the average store-bought type. Although this intuitively made a lot of sense to us, we never saw any proof that this claim was true, until *Mother Earth News* looked into the matter. They performed nutritional analysis on a number of small-flock, pasture-raised eggs that had been submitted to them by their readers and compared these results to published nutritional analysis of regular eggs. What they found were significant differences in cholesterol, fat balance, and vitamin content that favor the small flocks. Hurrah! Our suspicions were confirmed.

Eggs from pasture-raised flocks contain

- ⅓ less cholesterol
- ¼ less saturated fat
- ⅔ more vitamin A
- 2 times more omega-3 fatty acids
- 3 times more vitamin E
- 7 times more beta carotene

Source: Cheryl Long and Tabitha Alterman, "Meet Real Free-Range Eggs," *Mother Earth News*, October/November 2007, http://www.motherearthnews.com/Real-Food/2007-10-01/Tests-Reveal-Healthier-Eggs.aspx.

The survey suggested that, to gain these nutritional benefits, laying hens must be allowed to feed on pasture. In a small farm setting, this is accomplished by using mobile pens to safely move the birds around on grassy fields. At home, you will be doing something similar by allowing your birds access to your lawn and garden to graze and forage. Unfortunately, many home flocks are confined for most or all of the day to a small run where they have little opportunity to forage and graze. Fear not; these birds can still benefit from a diverse, plant-rich diet. As we mentioned earlier, if you can't bring the birds to the greens, bring the greens to them. Our customers have successfully supplemented their hens' diet with wilted greens from the garden or grocery store; fresh, short lawn clippings; and many other yummy items. Feel free to experiment a little; just be careful not to feed long blades of grass, dry grass, or seed heads that can become stuck in the birds' digestive system. Also, keep in mind that pasture or lawn feeding is no substitute for feed—it should constitute a portion of their diet only. The general rule of thumb is that a chicken should get two-thirds of its calories from feed.

There's at least one thing scientists can agree on about eggs: they contain the best-quality protein of any common food item. By quality, we are referring to the balance of amino acids (building blocks of proteins) in relation to usability by our bodies. That's right—eggs are more useful to your body as a source of protein than even a thick steak.

There's another thing that most scientists can agree on: eggs are delicious! Let's see what's cooking in our kitchen.

RECIPES

Everything you make with your homegrown eggs will be different from the versions you know made with store-bought eggs—and better! Whether your dish is sweet or savory, it will have a much richer flavor and color than if you had made it with store-bought eggs. You will discover that your dishes will look quite different with your backyard eggs in them. Hannah found that it was quite a conversation starter when she brought eggy dishes to potlucks. The quiche that everyone is used to being an anemic yellow was instead a vibrant orange. The flavor was out of this world. Unfortunately, she had to admit that the eggs deserved much of the credit, rather than her cooking prowess alone.

Over the years we and our families have collected a number of favorite egg recipes, and we would like to share some of them with you here. A truly fresh egg is an inspirational ingredient, and you will want to know how to put it to good use.

Hannah's Flan

This is a treat that we each used to seek out on restaurant menus, until we realized that we could easily make it ourselves. Since then, we have become hopelessly snobby about flan and tend to be disappointed in the pale, flavorless versions we get if we order it while out. The recipe uses plenty of eggs, is surprisingly easy to make, and gets gobbled up so fast that we always make it in large batches. Caramelized sugar is poured into individual ramekins prior to baking and magically becomes a delicious sauce for these custards after you remove them from the molds. Be sure to use the hot water bath—it seems like a hassle, but you will have seriously disappointing, rubbery flan without it.

Note: We think whole milk is tastiest, but the custard will also set with low-fat milk. Hannah likes to sprinkle some fresh nutmeg onto the top of each custard; another, more traditional option is cinnamon. ***Makes 8 servings***

3/4 cup plus 2/3 cup granulated sugar

6 eggs

3 cups whole milk

1 1/2 teaspoons vanilla

1/4 teaspoon salt

Nutmeg or cinnamon, for sprinkling (optional)

1. Preheat oven to 325°F.

2. Set out 8 ramekins and start boiling several quarts of water in a large pot—you'll need it soon. Caramelize the 3/4 cup sugar by pouring it all at once into a heavy stainless-steel frying pan and heating it over medium-high heat. Resist the temptation to stir it until it starts to melt and brown slightly around the edges. At this point, turn the heat down to low and stir with a wooden spoon only as much as needed to redistribute the sugar for even melting. Once the sugar crystals have melted and it has taken on a light caramel-brown color, remove the pan from the heat and pour the molten sugar to coat the bottom of each ramekin. Let the sugar cool and harden while you prepare the custard.

3. Beat the eggs in a large bowl until quite well combined, but avoid beating them into a froth, which will only cause unsightly bubbles. Add the milk, the remaining ⅔ cup sugar, vanilla, and salt. Place the ramekins in one or two baking dishes (depending on the sizes you have on hand) and pour an even quantity of the custard mixture into each cup. Sprinkle with nutmeg or cinnamon. Carefully pour the boiling water into the baking dish(es) around the ramekins as high as you can without overflowing the dishes or risking spillage when you put them into the oven.

4. Bake until a knife inserted into the center of one of the flans comes out clean, 50 to 60 minutes. Test the flans before removing from the oven. If you find that they are not quite done yet, check again in a few minutes. Try not to overcook.

5. Remove from the oven and carefully lift the ramekins out of the water bath. Set on a wire rack to cool to room temperature before serving. To serve, loosen the flan from the ramekin by running a knife around the edge. Then place a plate over the top and quickly flip it over. The flan will slip onto the plate, and the caramel sauce will pour over it.

Sunday Morning Crepes

These are something to look forward to on the weekends, when you have time to prepare them. Crepes are really quite easy, but require some forethought. We usually eat them as if they were pancakes, with maple syrup or homemade jam, but they can also be served rolled up around all sorts of delicious sweet or savory fillings. Make the batter the night before or at least 2 hours before cooking to allow the flour to soak up the liquid. Otherwise your crepes will be impossibly brittle and difficult to flip. We like to use all-purpose white whole wheat flour, which provides a slightly richer flavor and makes these seem slightly more virtuous. We also recommend whole milk, which is less virtuous but much more tasty. Double the recipe if you have a big crowd, would like extras to freeze, or just want to eat yourself silly.

Note: If you are using an electric crepe maker, you will find this batter too thin for your needs—follow the manufacturer's instructions for an appropriately thick batter. ***Makes about 12 large crepes***

3 eggs
2/3 cup milk
2/3 cup water
1/4 teaspoon salt
2 tablespoons canola oil
1 cup flour
Topping of your choice

1. Break the eggs into a blender and give them a quick spin to combine yolks and whites. Then add the milk, water, salt, and oil and blend until well mixed. Add the flour, a little at a time, until you have a smooth batter about the consistency of heavy cream. Cover and put into the refrigerator to rest. When you are ready to cook, give the batter a quick stir—it will have separated somewhat while resting.

2. Heat a nonstick pan over medium-high heat and use a cooking spray or smear the pan with a little oil. The pan is hot enough when a drop of water sizzles on it. Pour in a spoonful of batter and quickly tip the pan to distribute the batter evenly. You have used too much if the crepe is heavy and thick, and too little if your crepe looks like a spider's web. Cook the crepe for about 30 seconds, or until the edges start to brown slightly, then flip it over with a spatula—or with a flick of the pan if you have fancy skills. Cook the other side for another 20 seconds, then remove and serve immediately or fold into quarters and keep on a plate in a warm oven until all your crepes are cooked.

3. Don't be discouraged if your first crepe is a little ugly. In our house, this is the best excuse to sample the product and make sure it is satisfactory. As your pan gets warmer, and you get the knack, your crepes will become prettier and prettier.

Perfect Hard-Boiled Egg

Eggs provide a source of perfectly balanced proteins. Hard-boiled eggs are ideal packed away for snacks on the go, or incorporated into other recipes such as deviled eggs or egg salad. This is a great way to catch up if your chickens have been laying faster than you can eat, and you have a carton of eggs that are now a couple of weeks old. Slightly "aged" eggs produce boiled eggs that are much easier to peel. There are many approaches to boiling an egg—we recommend one of the simplest and most time-honored.

Note: The recipe is adjusted for refrigerated eggs. If you are using room-temperature eggs, reduce the cooking time by 2 minutes.

1. Place whole eggs in a single layer on the bottom of a pan and add cold water to cover by about an inch.

2. Set over medium heat until the water simmers.

3. Reduce heat to low and cook for an additional 15 minutes.

4. Immediately run cold water over the cooked eggs to prevent overcooking.

Perfect Poached Eggs

Warm eggs directly from the laying box will make the best poached eggs you have ever eaten. Not only will the flavor be out of this world, but the freshness helps the eggs hold their shape in the water and solves the most frustrating problem folks encounter in trying to poach an egg—egg-white spread. Poached eggs are the basis for any number of fancy egg dishes, but our favorite is Eggs Benedict—more on that later! Your goal is an egg with a firm, compact white and a thick but still liquid yolk. Strictly speaking, the egg yolk is not fully cooked at this point, but a fresh egg is extremely unlikely to make you ill, and no self-respecting chef would serve a poached egg with a hard-cooked yolk. (If your taste demands a firmer yolk, cook for a minute or so longer than instructed below.) The amounts of vinegar and salt specified are for 4 cups of water and 4 eggs. You may adjust them up or down if cooking significantly more or fewer eggs. ***Makes 4 poached eggs***

4 eggs
1 tablespoon white vinegar
1 teaspoon salt

1. Add one cup of water for each egg you intend to poach to a large, deep, preferably nonstick frying pan and heat until vigorously boiling.

2. Add the vinegar and salt to the boiling water and reduce to a simmer.

3. Crack an egg into a small bowl and slowly release it into the simmering water. Quickly repeat this process until all the eggs you intend to cook are in the water.

4. Cover and cook for 2 to 3 minutes, making sure the water stays at a mellow simmer. Remove the eggs in the order in which they were added with a slotted spoon and drain fully before serving.

Dale's Deviled Eggs

Robert's stepdad Dale makes deviled eggs that are out of this world. Luckily for all of us, he has generously agreed to share his recipe for the book. The secret is in the garlic, Robert's all-time favorite seasoning. ***Makes 12 deviled eggs***

6 boiled eggs, peeled
2 tablespoons mayonnaise
1 teaspoon french mustard
1 small to medium clove garlic, finely chopped
1 tablespoon finely chopped sweet onion (such as Walla Walla or Vidalia)
Salt and freshly ground black pepper
Paprika for sprinkling

1. Cut the boiled eggs in half lengthwise. Gently remove the yolks and place them in a separate bowl.

2. Add the mayonnaise, mustard, garlic, and onion to the yolks and mash with a fork until smooth. Add salt and pepper to taste.

3. Spoon the yolk mixture back into the empty whites and lightly sprinkle each with paprika. Refrigerate until you are ready to serve them.

Frittata

This is an easy egg dish that can be served as part of nearly any meal and will serve a number of people all at once. In general, fillings that would be delicious in an omelet will also work in this thicker, Italian version of the same. You'll need a pan that is nonstick but oven-safe. A seasoned cast-iron skillet will work well also. ***Makes 4 servings***

1 cup coarsely chopped onion
2 tablespoons olive oil
1 cup quartered baby red-skinned potatoes, boiled or steamed
1 cup coarsely chopped or sliced vegetables or cooked meats
1/4 cup fresh (or 1/8 cup dried) herbs of your choice
8 eggs, beaten
Salt and freshly ground black pepper
1/2 cup crumbled goat or grated mozzarella cheese
1/4 cup grated Parmesan cheese

1. Preheat the broiler. Sauté the onion in the oil over medium heat until translucent. Add the potatoes and continue to sauté until the potatoes are golden brown, about 8 minutes. Add the other vegetables, meats, and herbs and stir briefly to warm them.

2. Season the eggs with salt and pepper to taste. Pour the beaten eggs over the contents of the pan, then tilt the pan rapidly to distribute and evenly cook the egg, still over medium heat, until it is set, but the top is still quite moist, about 10 minutes.

3. Sprinkle the cheese over the top and place the pan on the top rack under the broiler just until the cheese is melted, about 3 minutes.

4. Sprinkle with the Parmesan, cut into wedges, and serve.

Hollandaise Sauce

This sauce is notoriously a beast to make, but it is so good that it is worth the effort. The key is to slowly and evenly heat the egg yolks while beating them vigorously to prevent separation. It helps to have all of your ingredients prepared in advance. Take those perfect poached eggs, plop them on top of an English muffin and Canadian bacon, drizzle with hollandaise sauce, and you have made your own Eggs Benedict. Hollandaise sauce is also a great topping for vegetables like asparagus.

Note: If you don't own a double boiler, use a combination of saucepan and a stainless steel mixing bowl that fits into it without its bottom touching the water in the pan. ***Makes 1 cup***

10 tablespoons clarified butter (ghee)
3 egg yolks
1 1/2 tablespoons cold water
2 to 3 teaspoons freshly squeezed lemon juice
Salt and freshly ground black pepper

1. Heat the butter over low heat in a small saucepan until warm. Meanwhile, heat about an inch of water in the bottom pan of a double boiler over high heat until barely simmering. Reduce heat to maintain a low simmer.

2. In the top bowl of the double boiler, vigorously beat the egg yolks and cold water together with a whisk until frothy. Place the bowl over the barely simmering water and keep whisking until the egg mixture is thickened and has increased in volume significantly, about 4 minutes.

3. Remove the bowl from the heat, continuing to whisk the eggs to cool them slightly. Slowly pour in the butter, whisking all the while. Add the lemon juice and salt and pepper to taste. If your sauce resembles a sauce rather than a curdled mess, you have succeeded! Serve immediately, or if you need a bit more time, keep the bowl warm over water that is no hotter than the sauce. If, despite your best efforts, the sauce separates, it can still be saved. Put an additional tablespoon of lemon juice in a separate bowl and whisk in a dollop of your sauce until smooth. Repeat dollop by dollop until the sauce is reincorporated. Disaster averted!

Chocolate Soufflé

Robert's favorite childhood dessert happens to use lots of eggs. We hope it turns out to be one of your favorites also. Soufflés are notorious for "falling," or collapsing inward instead of standing tall. The secret is not to remove it from the heat of the oven until you know for sure it is done. This involves testing for doneness while the soufflé is still in the oven, and doing so quickly. (And if your soufflé does fall, it will still taste delicious.) Follow this recipe, and we think you will be pleased with how easy it is to make. ***Makes 6 servings***

4 eggs
2 tablespoons unsalted butter
3 tablespoons all-purpose flour
3/4 cup 2 percent milk
1/2 cup high-quality semisweet chocolate chips
1 teaspoon vanilla
1/4 cup sugar, plus more for preparing the soufflé dish
Whipped cream

1. Preheat oven to 350°F.

2. Prepare a 2-quart soufflé dish (which looks like an oversized classic ramekin) by greasing and then sprinkling the sides with sugar. Separate the eggs, placing the yolks in one large bowl and the whites in another. Beat the yolks, but leave the whites as they are for now.

3. Melt the butter in a small saucepan over medium-low heat, then stir in the flour. Pour in all the milk and cook until bubbling and thickened, stirring all the while. Add the chocolate and continue to stir until completely melted. Remove from heat and gradually stir into the egg yolks.

4. Beat the egg whites and vanilla together until softly curling peaks form when you lift your whisk or mixer attachment straight up. Slowly add the sugar, whisking all the while until peaks no longer curl but stand up tall.

5. Fold half of the whites into the chocolate mixture to lighten it. Then fold the chocolate mixture back into the remaining egg whites just until fully incorporated. Avoid overmixing. Scoop gently into the prepared dish.

6. Bake for about 40 minutes. When done, a clean knife inserted into the middle of the soufflé should come out clean. If it is not done, close the oven quickly to prevent collapse, and cook another 5 minutes or until done.

7. Proudly carry your lofty soufflé to the table (or, if it has deflated, portion it out in the kitchen). Scoop into individual serving dishes, top with whipped cream, and serve immediately.

OVER EASY

We hope that these recipes have whetted your appetite for homegrown eggs. Likewise, we trust that this book has heightened your excitement for the rewarding hobby of chicken keeping. Whether it's the warmth of a freshly harvested egg, the look of delight in a child's eyes as she gently holds a chick, or the comical hopping of a hen reaching for a blueberry just out of reach, there's no doubt that you'll have an "Aha!" moment when you realize how worthwhile all your hours of poultry preparation have been. If you ever find yourself in Portland, Oregon, please stop by the store and let us know about your own chicken-keeping experience.

RESOURCES

Although we set out to make this book as complete as possible, there is much more to be explored. In addition to consulting other texts and friends with chickens, we find the Internet to be a great place to find information. We provide current information on this book's website: www.achickenineveryyard.com.

General Chicken Information and Products

Our Urban Farm Store's website features feed and care information, press clippings, our newsletter and blog, and more: www.urbanfarmstore.com.

This site is a must visit for all things backyard chicken (use the forum to ask questions): www.backyardchickens.com.

The City Chicken is another fine resource: http://home.centurytel.net/thecitychicken/index.html.

This fun page has a little of everything: www.chickencrossing.org.

This new site features a well-organized forum and listing of resources across the country: http://urbanchickens.org.

Foy's Pet Supplies is the oldest and one of the largest suppliers of specialty products for chickens: www.foyspigeonsupplies.com.

This is a well-designed and useful site: www.chickenkeeping.com.

Here's a nice site for education: http://urbanext.illinois.edu/eggs.

A quality blog: www.urbanchickens.net.

Our local chicken forum: http://groups.yahoo.com/group/PDXBackyardChix/join.

Backyard Poultry is probably the best known magazine on this topic and our personal favorite.

Urban Chicken-Keeping Ordinances and Laws

This site is the best list we've seen: http://home.centurytel.net/thecitychicken/chickenlaws.html.

Urban Chickens covers the basics: http://urbanchickens.org/chicken-ordinances-and-laws.

Breed Selection and Hatcheries

The one and only, complete chicken breed chart (click the initials under the breed name to link to sites with pictures): www.ithaca.edu/staff/jhenderson/chooks/chooks.html.

Murray McMurray Hatchery is one of the oldest and best-known hatcheries; chicken lovers pore over this site researching breeds and fantasizing about adding to their flocks: www.mcmurrayhatchery.com.

A relative newcomer to the scene, My Pet Chicken is known for shipping small numbers of birds in special containers—they also have great breed information: www.mypetchicken.com.

Scroll to near the bottom of this page for one of the most encyclopedic reviews of breeds (with lots of pictures and extensive links, too): www.feathersite.com/Poultry/BRKPoultryPage.html.

The American Poultry Association set the standards for breed perfection (if you're into that sort of thing): www.amerpoultryassn.com.

The Poultry Club fills a similar role in the UK: www.poultryclub.org/home.htm.

The American Livestock Breeds Conservancy helps to preserve heritage breeds: www.albc-usa.org.

So does Sand Hill Preservation Center: www.sandhillpreservation.com.

Poultry Health and Feed

A scientific, yet digestible, exploration of the subject of feeding poultry: http://attra.ncat.org/attra-pub/feeding.html.

A comprehensive and free online veterinary manual: www.merckvetmanual.org/mvm/index.jsp.

This site is a wealth of information about feed: www.lionsgrip.com/chickens.html.

This site will help you find an avian (bird) vet in your area: www.lafeber.com/findalocal/vet.

This is the official "find an avian vet" search engine from the Association of Avian Veterinarians: www.aav.org/search.

This quirky site has a nice listing of health links: www.poultryhelp.com/link-anatomy.html.

Coop Resources

Eglu is the original mod coop for suburban chicken keepers: www.omlet.us/homepage/homepage.php.

The Garden Coop offers detailed plans to help you build a modern and secure coop and run: www.thegardencoop.com.

Mother Earth News has a guide for converting a doghouse into a coop here: www.motherearthnews.com/Do-It-Yourself/2007-04-01/Portable-Chicken-Mini-coop-Plan.aspx.

My Pet Chicken has a number of coop plans and prebuilt coops for sale: www.mypetchicken.com.

This site has coops for sale that are inexpensive in comparison to those of other companies, as they are manufactured overseas. The designs seem well thought out: www.wholesalechickencoops.com.

Egg Information

The "Incredible Edible Egg" website by the American Egg Board has numerous egg facts, recipes, and info: www.aeb.org.

Go to this website and search on "egg washing" to find a great PDF document that details how to wash eggs without a sanitizer: www.ext.vt.edu.

The Maine Organic Farmers and Gardener's Association has information on how to wash and sanitize eggs. Search on "egg washing." The instructions are for hatching eggs, but are also appropriate for eggs to be consumed: www.mofga.org.

The USDA discusses proper egg handling on this website: www.fsis.usda.gov/factsheets/Focus_On_Shell_Eggs.

ABOUT THE AUTHORS

Together, ROBERT AND HANNAH LITT have kept and enjoyed chickens for almost ten years. After returning to school for a master's degree in agriculture, Robert became interested in animal nutrition and the local food movement. Dissatisfied with the chicken feed available to him at the time, he codeveloped a locally sourced, high-quality poultry feed with a nearby mill and sold it from his garage. At Hannah's urging, he opened the Urban Farm Store in early 2009 to sell the feed, along with chicks, edible plants, beekeeping equipment, and other homesteading items.

Robert continues to work at the store and teach aspiring chicken keepers the ropes, while Hannah focuses on her career as a nurse midwife and moonlights at the store. The couple has been featured on Planet Green's *Renovation Nation*, National Public Radio, Oregon Public Broadcasting, and KATU News, and Robert was recently named to *Food and Wine*'s "40 Big Food Thinkers Under 40."

Robert and Hannah live with their young daughter, Abigail, and flock (Rosie, Yankee, Wissahickon, Ginger, Butternut, Buttercup, and Tweedy) in Portland, Oregon. Visit www.urbanfarmstore.com and www.achickenineveryyard.com.

INDEX

CIGARETTES.
2
3

OGDEN'S CIGARETTES.
FOSTER MOTHERS.

OGDEN'S CIGARETTES.
WIND AND SUN
SCREENS FOR POULTRY.

OGDEN'S CIGARETTES.

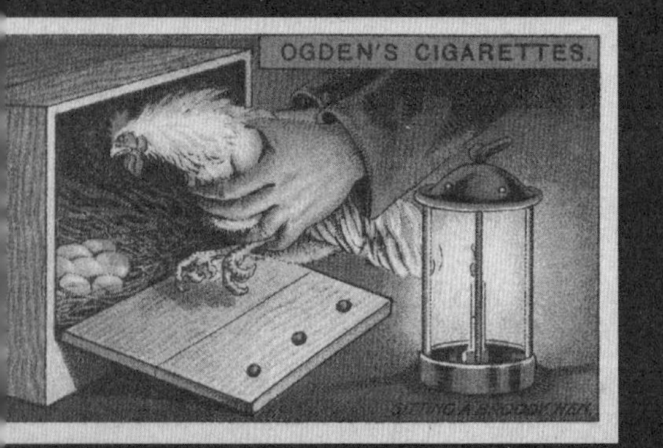
OGDEN'S CIGARETTES.
SITTING A BROODY HEN.

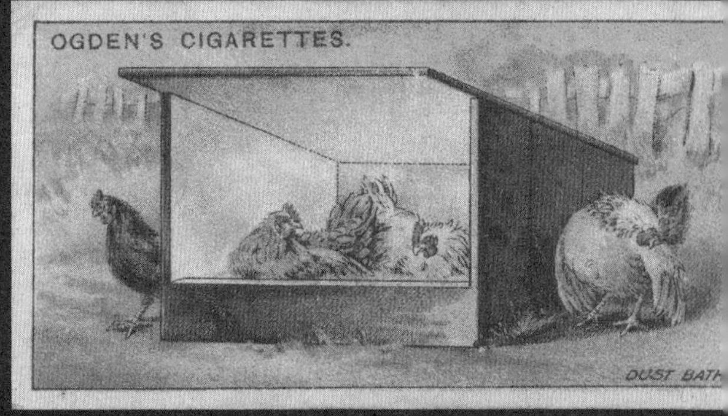
OGDEN'S CIGARETTES.
DUST

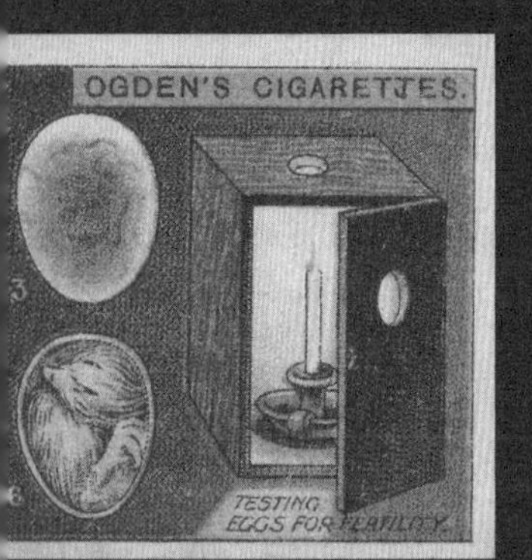
OGDEN'S CIGARETTES.
TESTING
EGGS FOR FERTILITY.

OGDEN'S CIGARETTES.
HOPPER FEEDING.